Darwin's Origin of Species

A Biography

Current and forthcoming titles in the
Books That Changed the World series:

The Bible by Karen Armstrong
Machiavelli's *The Prince* by Philip Bobbitt
Plato's *Republic* by Simon Blackburn
Thomas Paine's *Rights of Man* by Christopher Hitchens
The Qur'an by Bruce Lawrence
Homer's *The Iliad and the Odyssey* by Alberto Manguel
On The Wealth of Nations by P. J. O'Rourke
Clausewitz's *On War* by Hew Strachan
Marx's *Das Kapital* by Francis Wheen

Darwin's
Origin of Species
A Biography

JANET BROWNE

Grove Press
New York

First Published in Great Britain in hardback in 2006 by Atlantic Books

Printed in the United States of America

Library of Congress Cataloging-in-Publication Data
Browne, E. J. (E. Janet), 1950–
Darwin's Origin of species : biography / Janet Browne.
p. cm.
Includes bibliographical references and index.
ISBN-10: 0-8021-4346-6
ISBN-13: 978-0-8021-4346-4
1. Darwin, Charles, 1809–1882. 2. Naturalist—England—Biography.
3. Darwin, Charles, 1809–1882. On the origin of species. I. Title.

QH31.D2B843 2007
576.8′2092—dc22
[B] 2006043623

Grove Press
an imprint of Grove/Atlantic, Inc.
841 Broadway
New York, NY 10003

Distributed by Publishers Group West

www.groveatlantic.com

08 09 10 11 12 10 9 8 7 6 5 4 3 2 1

CONTENTS

ACKNOWLEDGEMENTS

Writing this book was a very enjoyable process and I am particularly grateful to my editor Louisa Joyner for her encouragement and support. The rest of the team at Atlantic Books were also fabulously efficient and friendly in seeing it through production. Jane Robertson worked wonders on my prose. Elsewhere, friends at the Wellcome Trust Centre for the History of Medicine at University College London have offered much useful advice. Special thanks, as always, are due to Bill Bynum and Michael Neve, very knowledgeable and stimulating Darwinian colleagues. I am also extremely grateful to the students who have, over the years, patiently discussed Darwin with me. This short study is written with them in mind. Most of all, this book is for Kit and Evie, students of other subjects, but only too familiar with Darwin over the dinner table. Their opinions are important to me and I hope this will provide a more connected story.

A NOTE ON EDITIONS

Charles Darwin's *On the Origin of Species by Means of Natural Selection, or the Preservation of Favoured Races in the Struggle for Life* was published in November 1859 in London by the firm of John Murray. The publisher's advertisements indicate that the most likely date of publication was Thursday, 24 November. This first edition is nowadays mostly seen only in rare book collections. Several modern reprints of the first edition text are available in different formats, including on the internet. The first edition has also been reproduced in the twentieth century as an exact photo-facsimile, the most well known being edited and introduced by the biologist Ernst Mayr and published by Harvard University Press in 1959. All quotes in the present volume, unless otherwise indicated, are taken from this facsimile.

The second edition was produced very soon after the first, on 7 January 1860. Darwin managed to make a few significant corrections. Three thousand copies were printed, making this the largest edition issued in Darwin's lifetime. Six editions were published by the time of his death in 1882, each one with corrections and alterations. The third edition (1861) is interesting because Darwin added a short 'Historical Sketch' in which he described other evolutionary theories. In the fifth edition (1869) he first used the expression 'survival of the fittest'. The sixth edition, issued in 1872, is usually regarded as the last that Darwin corrected. He intended it to be a popular edition. It was printed in smaller type and cost much less. It was extensively revised and included a whole new chapter in which he answered criticisms. Most modern copies of the *Origin of Species* are based on this edition.

At the same time, editions were published by Appleton in New York. These do not completely match the English ones in content because Darwin often supplied corrections and other material either in advance or after each London edition. Translations were issued in eleven different languages during Darwin's lifetime and he tried to supervise each one, not always successfully. The first French and German translations did not satisfy him and he sought out new translators, hence later editions in those languages are closer to Darwin's original intentions. The book has received detailed bibliographical attention from Richard Freeman in *The Works of Charles Darwin: An Annotated Bibliographical Handlist* (2nd edn, Folkestone, Dawson Archon Books, 1977). A sentence-by-sentence analysis covering the changes made to all editions in English in Darwin's lifetime was published by Morse Peckham, *The Origin of Species: A Variorum Text* (Philadelphia, University of Pennsylvania Press, 1959).

INTRODUCTION

Charles Darwin's *Origin of Species* is surely one of the greatest scientific books ever written. Yet it does not fit the usual stereotype of what we nowadays expect science to be. It is wonderfully personal in style. It has no graphs or maths, no reference to white-coated figures in a laboratory, no specialized language. The years leading up to its publication were crammed with unexpected setbacks, chance encounters, high emotion and controversy. It sold out to the book trade on publication day and the arguments that it ignited spread like wildfire in the public domain, becoming the first truly international scientific debate in history. Readers attacked it or praised it, and struggled to align their deep-seated religious beliefs with Darwin's disquieting new ideas. From the start it was acknowledged as an outstanding contribution to the intellectual landscape, broad in scope, full of insight and packed with evidence to support his suggestions – but passionately criticized at the same time for proposing that all living beings originated through entirely natural processes. Apes or angels, Darwin or the Bible, were burning topics for Victorians. Many of these issues are still very much alive today. In fact, the writing and controversial reception of Darwin's *Origin* were never set apart in some cold esoteric world of science. Its story, in many ways, is the story of the modern world.

From today's perspective, of course, Darwin's role as one of the makers of present times has never been more evident. His writings challenged everything that had previously been thought about living beings and became a crucial factor in the intellectual, social and religious transformations that took place in the West during the nineteenth century. In time, Darwin grew to be one of the most famous scientists of his day, a Victorian celebrity whose work even in his own lifetime was regarded as a foundation stone for the modern world. Were we descended from apes? Must we give up the story of Adam and Eve and regard our purpose in this world as meaningless, little more than an animal existence? It was not just a question of arguing about the literal truth of the Bible. Few people, even then, believed in the Garden of Eden as a real place. Instead, Darwin seemed to be expelling the divine completely from the Western world, calling into doubt everything then believed about the human soul and our sense of morality. If humans were no longer answerable to God, their creator, were they free to do what they liked, without any moral constraint at all? 'Is it credible that a turnip strives to become a man?' enquired Samuel Wilberforce, Bishop of Oxford, in 1860. Darwin was popularly supposed to have assassinated the idea of God and once, jokingly, labelled himself the 'devil's chaplain'.

Retrospectively, it is common to label those stirring times as the 'Darwinian revolution'. The words usually come with a warning attached, for it is now clear that many of the themes addressed by Darwin were not new, either to him or to his readers. Even so, the label retains much of its meaning in the mind of the public. As so often happens, one man and one book have come to represent a larger transformation in thought. Yet the impact of evolutionary ideas has waxed and waned since Darwin's death, paradoxically sometimes at the same time. At the end of the nineteenth century and beginning of the twentieth, for example, when the evolutionary imperatives of competition and progress were expressed in the social sphere through imperial expansion, free enterprise and eugenic doctrines, and the words 'survival of the fittest' were on every lip, many biologists felt that the scientific side of Darwinism was utterly incompatible with early genetics. Paradoxically again, in the 1930s and 1940s, just when a number of avant-garde biologists hoped to produce a new 'evolutionary synthesis', there was strong support for rival systems based on environmentalist ideas of the inheritance of acquired characteristics. Meanwhile, the John Scopes 'monkey' trial in Dayton, Tennessee, in 1925, in which the fundamentalist politician William Jennings Bryan led the prosecution against a science teacher charged with illegally teaching the theory of evolution, and the agnostic lawyer Clarence Darrow the defence, has gone down as a watershed in the relations between science and religion. For a while it was against the law in Tennessee to teach evolution in schools.

At the start of the twenty-first century, Darwin's ideas have never been more prominent – although arguments are as heated as ever. Transformed by the modern understanding of heredity, and refined in a thousand different ways as knowledge marches onwards, the idea of natural selection is the cornerstone of most biological thinking across the globe. Palaeontologists trace mass extinctions and bursts of change in the fossil record, molecular studies throw light on the origins and diffusion of early mankind and genes are regarded as an essential key to human behaviour, even to the workings of the mind. Such views naturally generate intense debate. Criticisms are raised against sociobiology and the tendency to reduce everything down to the action of 'selfish' genes. Philosophers suggest that selection theory is an invalid form of knowledge, not capable of demonstrable proof. Ordinary people look at the rampant commercial competition and exploitative economic policies around them and wonder whether altruism was ever a basic human trait, whilst modern creationists vigorously challenge the arguments used to support evolution and demand equal time on the school curriculum for the Christian creation story. In a *New York Times* election poll taken in November 2004, 55 per cent of responders believed that God created human beings in their present form.

Darwin would recognize many of these developments. However he was no godless radical striving to overthrow everything he knew. In personal terms, he was a highly respectable figure, hardly the kind of man that might be imagined to publish such a far-reaching text. He was never imprisoned for heretical views like the Italian natural philosopher Galileo Galilei. English villagers did not burn straw effigies of him as they did for the political revolutionary Tom Paine. He was not accused of sacrilege by the ecclesiastical courts like Bishop Colenso was. There were no anti-Darwinian riots. Instead, he was buried in Westminster Abbey in London in 1882 as one of the nation's most revered scientific figures: 'the greatest Englishman since Newton' said *The Times*.

One notable feature of the so-called Darwinian revolution, indeed, is the way that the man at the centre of the storm was widely applauded in personal terms. Much of this can perhaps be associated with the rise of science as a leading feature of Victorian society. Much can also be linked with the spread of middle-class economic and political values through the era. Perhaps too, despite all the controversy, something must be said for Darwin's tendency to keep apart from the fray. He hated the cut and thrust of public disagreement even while accepting that science generally progresses through debate and argument. He much preferred to be a countryman, pottering around his garden in Kent. He liked to write letters, see friends and carry out small natural history experiments in his greenhouse or study. In some ways he could almost have stepped out of a novel by Anthony Trollope, a tall, quiet, likeable man with a modest, trustworthy air about him, deeply engaged with his work and family, committed to the idea of scientific truth. From time to time, like many

Victorian gentlemen, he was plagued by stomach trouble and mysterious disorders. A good Victorian paterfamilias, he grew a large beard, kept a careful eye on his investments and loved his wife and children. Somewhat surprisingly, he counted several vicars among his relatives and closest acquaintances. He enjoyed being a traveller, husband, father, friend and employer as well as a naturalist and thinker.

Above all else, he was indisputably an author. As an old man, looking back over the arguments that had surrounded him, he ruefully acknowledged the way in which *Origin of Species* had dominated the era. 'It is no doubt the chief work of my life,' he wrote in his *Autobiography*. 'It was from the first highly successful.'

Darwin's volume, under the full title *On the Origin of Species by Means of Natural Selection, or the Preservation of Favoured Races in the Struggle for Life,* was published in London by the firm of John Murray on 24 November 1859. It was a rather ordinary-looking volume bound in sturdy green publisher's cloth, 502 pages long, and somewhat expensively priced for the Victorian book market at 14 shillings – more than a week's wages for a labourer. The author's serious intent was obvious. There were no eye-catching natural history illustrations, no pedigree pigs or cows decorating the cover, not even a frontispiece of a prehistoric scene as there might be today in a book on evolution. The modest air suited its author perfectly. 'I am infinitely pleased & proud at the appearance of my child,' Darwin told Murray when an advance copy arrived in Yorkshire, where he was taking the water cure. 'I am so glad that you were so good as to undertake the publication of my book.'

These quiet words hid a wealth of previous drama and plenty of ferment to come.

CHAPTER 1

Beginnings

The history of *On the Origin of Species* began long before publication day.

Charles Robert Darwin was born in Shrewsbury in February 1809, the fifth child and second son of a prosperous medical doctor, Robert Waring Darwin, and his wife Susannah Wedgwood. The family took a leading role in respectable provincial society and were often to be found visiting relatives, participating in local charitable ventures or taking scenic holidays on the Welsh coast. Darwin recalled his earliest days as very happy ones even though his mother died when he was eight years old. In his *Autobiography* he said he had few recollections either of her or her death, perhaps because his three older sisters cared for him afterwards with great maternal affection. As far as can be discerned, this major event in his childhood left him with no conspicuous psychological problems. He appears to have been a warm-hearted boy who liked nothing better than to be with his friends and family, had a great love of the countryside, enjoyed reading a wide variety of books and listening to music. He was much loved in return, and all the available manuscripts remaining in libraries and archives around the world confirm that he grew up affable, talkative and cordial, despite all the illness and controversy to come, with the gift of making lasting friendships, and able to sustain a close and happy marriage until the end of his life.

One of his grandfathers was Erasmus Darwin, the poet, early evolutionary thinker and physician. The other was Josiah Wedgwood, the famous potter. Both made notable contributions to the Industrial Revolution and were key members of the extraordinary intellectual flowering of the eighteenth century. Such a remarkable family tree always excites comment and it has long been popular among historians to trace some of Darwin's personal ingenuity back to these two male figures in his ancestry. In reality, he bore no resemblance to either of them in character, except that he too was brought up in an intellectual, freethinking and scientific family atmosphere. By the third generation, however, the Wedgwood wealth made a considerable difference. This rather modern combination of manufacturing affluence, gentlemanly social standing, religious scepticism and cultivated background ensured that Darwin always had a place in upper middle-class society and the prospect of a comfortable inheritance, both of which served as very material factors in his later achievements. He was born, so to speak, into the financially secure intelligentsia of Britain.

Darwin attended Shrewsbury School (a private boys' school) from 1818 to 1825. As a boy he hoped to become a doctor and sometimes accompanied his father on medical rounds. He loved collecting natural history specimens. At school he enjoyed chemistry, and he and his older brother Erasmus set up a small laboratory at home for performing experiments during the holidays. These enthusiasms were relatively typical for young men of their social class and period; none the less they reveal the start of Darwin's lasting fascination with science and the natural world. Like many boys, he seemed content otherwise to wander about the countryside following his own interests. The documents preserved from those days suggest that he did not thrive in the rigid classical structure of male education at the time.

Life took an exciting twist when his father took him away from school early, and in 1825 sent him and his brother Erasmus to Edinburgh Medical School, where he began studying medicine. In those days students paid on an individual basis for whichever medical courses were necessary – anatomy, midwifery, physic, materia medica – a much more informal arrangement than today. Very young men could attend the university by taking a handful of courses before they settled down to serious study. After a diligent start, sixteen-year-old Darwin found the realities of early nineteenth-century medicine upsetting. Two 'very bad' operations, one on a child, convinced him he would never make a doctor (this was long before the age of anaesthetics) and he left in 1827.

During that short period, however, he was exposed to some of the most formative influences of his youth, influences that lasted right through to his death. Biographers regularly go back to Darwin's Edinburgh years, convinced that the seeds of all his later thinking lie there – and to a large degree they are right. Edinburgh University was the leading centre in Britain for science and medicine. It kept abreast of continental research and offered classes, both inside and outside the university, on all aspects of modern science. Darwin took Thomas Hope's chemistry class and Robert Jameson's natural history course, the latter supported by a fine natural history museum. He liked the museum very much. There he met a local taxidermist, a freed slave called John Edmonston, who had arrived in Scotland from the West Indies, who taught him how to stuff birds; and he spent pleasant hours talking with the curator William Macgillivray about shells and birds. Jameson's course introduced him to the subject of geology and he became aware of contemporary debates about the history of the earth and fossils – although he said he hated Jameson's dry and dusty lectures, and vowed never to pursue the subject again.

Darwin also enjoyed a lot of independent practical natural history work. He joined the Plinian Society, a small student society, where he met Robert Grant, a charismatic university lecturer in the medical school who approved of French developmental anatomy and evolutionary views. Under Grant's guidance Darwin began observing soft-bodied marine organisms from the North Sea, and made his first discovery in science, on the ova of Flustra, a kind of gelatinous 'sea-mat', that was announced at the Plinian Society on 27 March 1827. He found that the 'ova' were not eggs at all, but free-swimming larvae.

Grant dramatically broadened Darwin's perspectives. He took him into Edinburgh's scientific circles, and encouraged him to expand his natural history interests. From him Darwin acquired a lifelong fascination with 'generation' (sexual and asexual reproductive processes) and the embryology of invertebrates like molluscs, sponges and polyps. Grant encouraged Darwin to read Lamarck's *System of Invertebrate Animals* (1801) and one day burst out in praise of Lamarck's views on transmutation (sometimes also called transformism; the word 'evolution' was not used at that time). Darwin recalled that he listened, as far as he knew, with little effect on his mind. Yet he had already read his grandfather's book on the laws of life and health, the *Zoonomia* (1794–96) which included a short section setting out a theory of development very similar to Lamarck's. By then Erasmus Darwin and Lamarck had been dead for several decades but were by no means regarded as old-fashioned. They were highly valued by radical thinkers in the 1820s for their bold biological theories in the Enlightenment tradition, especially for their transmutationary ideas. Grant used these ideas, appropriately updated, to propose sponges as the basic organism from which all other forms developed to make the evolutionary tree. Darwin therefore left Edinburgh with much wider intellectual horizons than many young men of his age. He had already learned to see the value of lofty questions about origins and causes, and directly encountered evolutionary explanations for the patterns of life, although there is no reason to think that he became an evolutionist at that time.

Darwin's father was not pleased about his son's change of direction. After a few terse discussions at home, and extensive coaching in all the Latin and Greek that he had forgotten from school, Darwin shortly afterwards entered Christ's College, Cambridge, to read for an 'ordinary' degree, the usual start for taking holy orders in the Anglican Church. While the family was not particularly religious, entering the church as a vicar was an accepted route to a respectable middle-class profession in Victorian times and several members of the Darwin and Wedgwood circle were competent country parsons. Rather in the tradition of the Reverend Gilbert White, author of *The Natural History of Selborne*, young men with appropriate social and educational credentials could expect to find a comfortable niche in a country parish with plenty of time to pursue natural history or sporting interests. Darwin later said in his *Autobiography* that he was content with the idea of becoming a clergyman, though he felt one or two fleeting religious doubts. Afterwards he was well aware of the irony. 'Considering how fiercely I have been attacked by the orthodox it seems ludicrous that I once intended to be a clergyman'.[1] His father had evidently impressed on him the importance of gaining a profession: he could not depend on a full private income from inheritance alone. 'You care for nothing but shooting, dogs, and rat-catching, and you will be a disgrace to yourself and all your family,' Dr Darwin once declared, to his son's mortification. If not medicine, then the church seems to have been the likely subject of their conversations.

These years at Cambridge University were extremely significant for Darwin's later life, although not quite in the way that either Darwin or his father expected. Because of this, historians of science have combed through his experiences there, seeking even the smallest hint about his future preoccupations. All agree that the academic environment at Cambridge was very unlike Edinburgh and that the switchback ride from a coolly austere medical context to the lush theological pastures of Cambridge was decisive. Darwin's later achievements, in fact, can conveniently be characterized as a mix of Edinburgh and Cambridge ideas – the two traditions sparking insights off each other. At Cambridge Darwin entered the elite social and intellectual milieu he was to occupy for the rest of his days, and the friendships he made proved enduring. Of these, John Stevens Henslow (1796–1861), the young professor of botany, and Adam Sedgwick (1785–1873), an equally young professor of geology, were most important. He became acquainted with the scientist-philosopher William Whewell and the naturalist-parson Leonard Jenyns. His closest personal friend was his cousin William Darwin Fox, also at the university training to become an Anglican clergyman. For a couple of terms they shared rooms together, as well as some student debts and a dog.

Darwin had a fabulous three years. The lecture schedule was undemanding and there was plenty of time to indulge natural history interests. In the company of his cousin, Darwin became a passionate amateur entomologist, knowledgeable enough about beetle classification to send a minor contribution to the author of an authoritative textbook. He hunted foxes, shot game birds, swapped natural history specimens with his friends, played cards and enjoyed life with a wide circle of acquaintances. 'I got into a sporting set, including some dissipated low-minded young men. We used often to dine together in the evening, though these dinners included men of a higher stamp, and we sometimes drank too much, with jolly singing and playing at cards afterwards. I know that I ought to feel ashamed of days and evenings thus spent, but as some of my friends were very pleasant and we were all in the highest spirits, I cannot help looking back to these times with much pleasure.'[2]

On the academic side, as well as struggling through the required courses in mathematics, classics and theology, he went to Henslow's botany lectures and (in his final year) Sedgwick's public lecture course on geology. Henslow obviously liked Darwin very much – perhaps seeing something of promise in him – and began to invite him to evening parties where he could meet some of the eminent men of the university. On his advice, Darwin read widely, afterwards citing John Herschel's *Preliminary Discourse on the Study of Natural Philosophy* (1830) and Alexander von Humboldt's *Personal Narrative* (English translation 1814–29) as inspirational.

In particular Darwin engaged with the theological views of Archdeacon William Paley, initially as part of his syllabus and then as independent reading. Darwin was expected to be able to answer questions in the final examinations on Paley's *Evidences of Christianity* and *Moral Philosophy*. After he graduated, he read the last of Paley's trilogy, *Natural Theology* (1802), with its argument that the adaptation of living beings to their surroundings was so perfect that it proved the existence of God. How could such perfect design have come about, stated Paley, except from the careful hands of a designer? If a watch were accidentally found on a path, we would be entirely justified in thinking that it had been constructed by a skilled craftsman according to some design or plan. Such intricate mechanisms do not suddenly appear out of nothing, like magic. They are made by a maker. So, Paley argued, the world about us must be considered in the same way as the watch.

This natural theological standpoint dominated Cambridge teaching across the board, although not without criticism, and formed the cornerstone of Cambridge natural science. The Christian God, it was said, had created a world in which everything had its place and was designed to do its job properly – a point of view originally popular throughout the learned world in the sixteenth and seventeenth centuries, and which gained special support in Britain in the early nineteenth century. The physical world was thought to be governed by natural laws that ran like clockwork and even the underlying structure of society appeared to mirror a carefully regulated and well-designed piece of machinery. For many people at that time the image of God was not that of an absolute monarch, sending miracles and thunderbolts, but of a careful, all-seeing guardian who arranged everything to run efficiently. Natural theology, indeed, was commonly regarded by the British cultural establishment as one of the strongest bulwarks against social unrest because it reinforced ideas of stable hierarchy, a powerful antidote against civil insurgencies and rebellion. Theological doctrine, in this regard, was fully integrated into the political and social ethos of the most influential men of the early years of the century – the Cambridge network, as it has been characterized.

Paley's clear language gave Darwin great pleasure. 'The logic of this book [*Evidences of Christianity*] and as I may add of his *Natural Theology* gave me much delight... I was charmed and convinced by the long line of argumentation.'[3] Many of Darwin's later investigations of animal and plant adaptations were undertaken partly to provide an alternative to the perfect design described so eloquently by Paley. In a more literary, emotional sense, Paley also gave Darwin the words with which to express appreciation of the marvellous intricacy of natural beings, the glint of an insect wing or the small sacs of nectar at the base of flowers for bees to suck. While Darwin eventually discarded all notions of a designer-god, he always kept alive the sense of wonder that he had learned from Paley and never quite abandoned those early feelings of worship.

Second, it was Cambridge that handed him the future in the form of the *Beagle* voyage. All these youthful incidents and carefree developments would probably have come to nothing if Darwin had not gone on the long sea voyage that transformed his life. Initially, after his final examinations in 1831, he intended simply to enjoy himself until returning to

Cambridge in the autumn for theological training. Inspired by reading Humboldt's travels he wanted to make a natural history expedition to Tenerife with Henslow, but the logistics overwhelmed them and the plans never really got off the ground. So his other professorial friend, Adam Sedgwick, took Darwin as an assistant for two weeks on his summer fieldwork examining the earliest known rocks in Wales. Sedgwick taught him geology in the field and introduced him to the rationale for sound scientific decisions. These two weeks gave Darwin a lifelong love for geological theorizing on a large scale. He then went to his uncle's country house for the August shooting.

On his return to Shrewsbury, Darwin found a letter from Henslow offering him a voyage around the world on a British surveying ship, HMS *Beagle*. The invitation had come through several hands and was very unusual, even in its day. It originated from Captain Robert FitzRoy (1805–85), who requested permission from the Hydrographer of the British Admiralty to take with him a gentleman who could make good use of the journey for collecting natural history specimens. Such a gentleman would share the captain's facilities as a guest and was expected to pay his own way. The elite social network that linked government, naval administration and the old universities had led to a number of Cambridge professors being consulted – at one point Henslow himself thought he would like to go. So did Leonard Jenyns. But each felt that his parish commitments obliged him not to pursue it. As a result, Henslow thought Darwin was 'the very man they are in search of'.[4] It was not an official position, nor was it an offer to be the ship's naturalist, although in effect this did become the case. Robert FitzRoy was a young man himself, only four years older than Darwin, who was deeply interested in science and new developments in marine navigation. He believed that the voyage would offer a fine opportunity to advance British science.

At first Dr Darwin felt his son should not accept. The whole plan was 'a wild scheme' he declared. Disappointed, Darwin wrote down his father's objections. Prime among them was 'disreputable to my character as a clergyman hereafter... I should never settle down to a steady life... you should consider it as again changing my profession... that it would be a useless undertaking.'[5] Fortunately, Dr Darwin was persuaded otherwise by his brother-in-law, Josiah Wedgwood the second. The rest of the summer passed in a flurry of eager organization. 'The voyage of the *Beagle* has been by far the most important event in my life and has determined my whole career,' declared Darwin.[6] To the end of his days he would still thrill to the memory of that extraordinary experience.

Today the fame of this voyage sometimes makes it hard to remember that its purpose was not to take Darwin around the world but to carry out British Admiralty instructions. The ship had been commissioned to complete and extend an earlier hydrographical survey of South American waters that had taken place from 1825 to 1830. FitzRoy had joined the *Beagle* two years into that former voyage. The area was significant to the government for commercial, nationalistic and naval reasons, buttressed by the Admiralty's marked enthusiasm for practical scientific advance and preoccupation with accurate naval charts and safe harbours. In fact the Hydrographer's Office was renowned for sending out a great many surveying expeditions in the lull after the Napoleonic wars to promote and exploit British interests overseas. FitzRoy's interest in science encouraged him to equip the ship for its second voyage with several sophisticated instruments and a number of chronometers for taking longitude measurements around the globe. The voyage lasted from December 1831 to October 1836, during which time they visited the Cape de Verd Islands, the Falkland Islands, many coastal locations in South America, including Rio de Janeiro, Buenos Aires, Tierra del Fuego, Valparaiso and the island of Chiloé, followed by the Galápagos Islands, Tahiti, New Zealand, very briefly Australia and Tasmania, and the Keeling (Cocos) Islands in the Indian Ocean, concluding with the Cape of Good Hope, St Helena and Ascension Island. Darwin made several long inland expeditions on his own in South America, including a tour across the Andes. Whenever possible, he arranged with FitzRoy to be dropped off and picked up at various points.

The high profile of the voyage has sometimes also led to Captain Robert FitzRoy being harshly misrepresented. He was hardly the bible-waving caricature that is usually described in the literature. Admittedly there is a poignant symbolism in these two men travelling the world together, the one a sincere religious believer, the other *en route* to destroying the presence of God in nature. Yet at that time FitzRoy was a keen amateur geologist with rather advanced non-biblical views. He gave Darwin the first volume of Charles Lyell's landmark volume, *The Principles of Geology* (1830–33) and discussed with him some of the theories it contained. Darwin received the other two volumes during the voyage. It was only afterwards that FitzRoy became a pronounced biblical fundamentalist. There is no evidence that they disagreed about religion on board ship, although it is evident from their writings that personal relations were sometimes strained. They argued, a couple of times very intensely, but the arguments were about each other's manners not religion. On the whole they managed very well. Darwin usually ate with the captain and talked about all kinds of things with him as a friend, while sharing a cabin and workspace with two junior officers, the mate and assistant surveyor John Lort Stokes and fourteen-year-old midshipman Philip Gidley King. On the way home, Darwin and FitzRoy jointly wrote a short newspaper article praising the work of the Anglican missionaries on Tahiti. Images of Darwin alone with his thoughts on board the *Beagle*, arguing about religion with the captain, a solitary naturalist voyaging through strange seas of thought, is attractive but only true in part.

These five years on the *Beagle* voyage were the making of him. Some of them were spent galloping around on hired horses, striking camp in new places every night, hunting game for supper with companions from the ship, discussing the news from back home and enjoying himself; they were an extension of the carefree days as a Cambridge undergraduate. It seems very likely, in fact, that he was chosen for the voyage partly because of his cheerful ability to join in with the ship's activities, which combined pleasantly with his cultivated background and skill in shooting and hunting. There were plenty of occasions to display such attributes. In Montevideo the *Beagle* men marched into town armed to the teeth to quell a political uprising. In Tasmania they attended a very fine concert. In the far south they were nearly capsized by a calving glacier. Out in the forest near Concepción Darwin felt the earth buckle under his feet in a major earthquake. He swam in coral lagoons, was entranced by birdsong in a tropical forest, and contemplated the stars from the top of a pass on the Cordillera de los Andes. In Brazil, his passionate heart burned with indignation about slavery, still a legal system under Portuguese rule, and he listed some terrible tales in his diary: facts so revolting, he said, that if he had heard them in England he would have thought them made up for journalistic effect.

Throughout, he displayed an enthusiasm that FitzRoy and the other officers described as very engaging. They nicknamed him 'Philos', standing for 'Ship's Philosopher', or sometimes 'Flycatcher', and teased him about the natural history rubble that he accumulated on deck. For the full five years he remained good-natured and easy to get along with – a considerable feat on a small ship crammed with seventy-four men and boys. Only sea-sickness set him back. Darwin constantly fell sick when the ship was underway and never gained any sea legs. The captain and his cabin-mates were very sympathetic.

He was also free to explore every ramification of his love for natural history. Darwin took his responsibilities seriously. He made collections of birds, vertebrates, invertebrates, marine organisms, insects, fossils and rock samples, and a fair collection of plants. These were regularly shipped back to Henslow in Cambridge, who kept hold of them until his return. It was a good collection, including many unusual and new species, but it is still useful to note that it is probably only Darwin's subsequent fame that has made these animals and plants such important trophies in today's museums and institutions. In addition Darwin dissected and observed under the microscope that he kept in his cabin, keeping notes as he went along. All the time, he made extensive observations on habitat, behaviour, colouration, distribution and suchlike, creating a careful paper record that would form the basis of several books and articles after the voyage ended. He told his sisters and friends of the great satisfaction that these activities gave him. 'Looking backwards, I can now perceive how my love for science gradually preponderated over every other taste,'[7] he said late in life. During these years he trained himself to see – to look attentively at details – and to record. In retrospect, perhaps the most significant aspect of the voyage was therefore not the huge collection of specimens, the sights, the dangers, or even the personal maturation and friendships he experienced, but the opportunity to develop an intense understanding of the variety of the natural world. By the time he came back he had stopped shooting. 'I discovered, though unconsciously and insensibly, that the pleasure of observing and reasoning was a much higher one than of skill and sport.'[8] The impact of seeing so many different places and people and encountering such a variety of natural habitats and forms of life was incalculable. His eventual prominence as a naturalist ultimately rested on these long and careful days learning to observe and think about nature's prodigal abundance.

Darwin's mental development on the voyage should consequently be given its due. Any number of young men attended Grant's or Jameson's or Sedgwick's lectures, any number of enthusiasts collected natural history specimens. Few of them asked the kind of questions Darwin came to ask. Sometimes Darwin saw organisms that were excellently adapted to their way of life, just as William Paley had described. Sometimes they were very poorly 'designed'. Many of these issues were only fully explored after the ship returned in 1836. None the less, in the introduction to *Origin of Species* Darwin stated that three findings from the voyage were the starting point for all his views. These were the fossils he dug up in Patagonia, the geographical distribution patterns of the South American Rhea (ostrich), and the animal life of the Galápagos Archipelago.

The fossils were an extraordinary find. Located near Bahia Blanca (south of Buenos Aires), these remains of gigantic extinct mammals were later identified by London's museum experts as belonging to previously unknown species of Megatherium, Toxodon and Glyptodont. Darwin noted that the extinct animals were built on broadly the same anatomical plan as the current inhabitants of the pampas. There seemed to be a continuity of 'type' over long periods of time. Then, in the far south of present-day Argentina, he collected a species of Rhea (well known to the local inhabitants), which was smaller than the northern form. He liked to tell a funny story about this Rhea. The ship's company had caught a bird for the cooking-pot and it was not until it was half-eaten that Darwin realized that it was an unknown species he wanted for his collection. The bits that were left were later named *Rhea darwinii* in his honour (the name is now changed). He afterwards used the two kinds of Rhea to illustrate the fact that closely related species do not generally inhabit the same area – they are mutually exclusive. To his mind, it looked as if there might be some kind of family links either across time or in geographical space. He began to wonder why there should be such connections.

As the ship moved, so did Darwin's thoughts. In September 1835 the *Beagle* left South America and struck out for the Pacific, with its first call at the Galápagos Islands. Ironically, Darwin did not notice the diversification of species on the Galápagos Islands during the *Beagle*'s five-week visit, even though the English official on Charles Island (Isla Santa Maria) informed him that the giant tortoises were island-specific. However, everything about the islands impressed him greatly. He was fascinated by the iguanas that overran the land and seashore, giant tortoises, mocking birds and boobies, as well as the arid volcanic landscape and curious lichen-festooned trees. These fourteen tiny specks of land were right on the equator, swept by cold southern waters that brought fur seals and penguins to their shores. They were mostly within sight of each other but separated by deep, treacherous sea channels. The animals and birds were not used to human intruders and were very trusting in their behaviour. For the *Beagle* men it was almost like encountering a Garden of Eden. Darwin rode a tortoise, caught an iguana by its tail and came so close to a hawk that he could push it off the branch with his gun.

The birds that he collected were bundled together in a single bag: he never suspected that their individual location might be important. He did notice that the mocking birds seemed different from island to island, and were different again from those of continental South America. This observation was sufficiently perplexing for him to mention it in his ornithological notes some months later on the return voyage. He appears to have thought that the birds might be geographical varieties of one or more South American species – and reflected on the problem:

> When I see these islands in sight of each other, and possessed of but a scanty stock of animals, tenanted by these birds, but slightly differing in structure, and filling the same place in nature, I must suspect they are only varieties… If there is the

> slightest foundation for these remarks the zoology of archipelagos will be well worth examining; for such facts would undermine the stability of species.[9]

At Cape Town in June 1836 he may have discussed the creation of species by natural law with John Herschel, the great astronomer, at that time living in South Africa to observe the southern heavens. It is unlikely, however, that Herschel would have contemplated a natural origin for species. He had recently read Lyell's *Principles of Geology*. Herschel wrote to Lyell, whom he knew personally, to declare that the origin of species was a divine mystery, 'that mystery of mysteries' as Darwin later put it.

One other feature of the voyage proved of lasting significance although Darwin did not refer to this in the *Origin of Species*. His intellect was permanently stirred by the diverse human populations he met, and his *Beagle* writings contain colourful references to the gauchos, with whom he travelled across Argentina, Patagonian Indians, statuesque Tahitians, fierce Maoris and Australian aborigines as well as missionaries, colonists and slaves. Throughout he expressed the view that humans were all brothers under the skin. In fact a strong antipathy to slavery in any form was crucial to his developing views about the unity of all mankind. Anti-slavery politics were integral to his family viewpoint in general, for the first Erasmus Darwin had been an active promoter of emancipation causes in Britain and in his poems publicly praised Josiah Wedgwood's famous medal emblazoned with the motto 'Am I not a man and a brother'. Darwin's father, sisters and cousins all supported the anti-slavery movements of the early nineteenth century – as did he. And the *Beagle* was travelling the world just when these mass philanthropic movements reached their pinnacle in Britain with the Emancipation Act of 1832.

The only time that Darwin was really angry with Captain FitzRoy was over an incident at a great *estância* in Brazil, where the slave-owner called all his men before him and asked whether they wished to be free. No, they answered. Talking in the cabin afterwards, FitzRoy complacently took that response as a simple truth until Darwin pointed out that no slave would risk any words to the contrary. The captain stormed out of the cabin, saying that they could not live together any longer. On another occasion Darwin was given an insight into the slaves' attitudes: one day in Brazil, as he was being ferried across a river by a coloured boatman, he absentmindedly waved his arms to give directions and was horrified to see the man cower in fright because he thought he would be hit.

But Darwin's most unsettling encounter was with the indigenous inhabitants of Tierra del Fuego. He was deeply shocked by his first sight of them in their rudimentary wigwams, a canoe-going people who seemed to him to possess no resources whatsoever except the ability to make fires, after which the region was first named by Magellan. 'The sight of a naked savage in his native land is an event which can never be forgotten.'[10] The shock was all the more vivid when compared with three Anglicized Fuegians on board, who had been taken to England by FitzRoy on the previous *Beagle* voyage, educated by a clergyman, and were now being repatriated in a Protestant mission station that FitzRoy intended setting up close to their home territory in inland Tierra del Fuego. In London the three had quickly adopted European habits and speech. Now, Darwin was astounded at the difference between the Anglicized Fuegians and the indigenous tribes to which they belonged. 'I would not have believed how entire the difference between savage & civilized man is. It is greater than between a wild & domesticated animal.'[11] The fact that near-savages could be civilized (as Darwin saw it) confirmed his belief that, under the skin, humans were all one species. This belief remained a lifelong commitment. During the *Beagle*'s time in the far south Darwin and FitzRoy were disappointed to see that the three Anglicized Fuegians rapidly reverted to the aboriginal state. The trappings of civilization were only ephemeral, the two travellers mused.

Most important of all, however, was the attention Darwin paid to geology. He was delighted by the grand theoretical schemes he found in Charles Lyell's *Principles of Geology* and excited by Lyell's rejection of biblical authority as a source of geological explanation. The book was commonly regarded as theologically radical. Although Henslow had recommended that Darwin should read it, he also advised 'on no account to accept the views therein advocated'. What bothered Henslow – and what ultimately became so appealing to Darwin – was Lyell's insistence that the earth's changes were not necessarily progressive in nature. The surface of Lyell's earth was forever on the move, but the alterations were not directed by God towards any future point. At that time, few geologists believed that the earth had literally been created in six days. They saw the Bible more as a metaphor for the stages that the earth must have undergone from its beginnings to the present day. Yet most geologists connected this sequence with the broad outline of earth history indicated in the Judeo-Christian tradition – that is, that the earth was stocked by divine fiat and progressively shaped in six or seven stages by God for human habitation.

In his *Principles of Geology* Lyell challenged this view by claiming that the earth's surface showed no evidence of stages. Instead it constantly experiences innumerable, tiny, accumulative changes, the result of natural forces operating uniformly over immensely long periods. These changes were for the most part so small that they were usually unnoticeable to the human eye. But if they were repeated over many epochs they added up to substantial effects. He shocked his colleagues by insisting that the earth was immeasurably old, that it had no beginning and no vestige of an end, and would continue endlessly in never-ending geological cycles characterized by the successive elevation and subsidence of great blocks of land relative to the sea. There was no God-given direction or progression. The great Cambridge philosopher William Whewell, himself very interested in geology, dubbed this approach to the earth 'uniformitarianism'.

In Lyell's estimation geology also included what we now

call biology. He argued that there were no successive sets of animals and plants either, and that each species had been created in a piecemeal fashion, one by one. By saying so, he found himself in the middle of a logical dilemma. Gradualism in geology implied gradualism in biology – if the rocks slowly transform in a seamless web of change, then so could animals and plants. But since Lyell was not prepared to believe in any kind of transmutation in living beings, he quickly fell into a tangle of equivocation. In order to demonstrate that he did not believe in evolutionary matters, he provided a long and aggressive attack on Lamarck. All the available evidence indicates that Darwin read this attack with mounting interest: although the words were negative he was exposed to evolutionary information that was to play a significant part in his intellectual growth. From Robert Grant's enthusiasm in Edinburgh to Charles Lyell's opposition in Patagonia, Darwin recognized the passion that transmutation inspired – and the hostility.

Darwin went on to absorb Lyell's teachings, using his geological ideas to explain the landforms he saw; they supplied the groundwork of his three later books on the geology of South America. Here and there he audaciously produced explanations for geological structures that he thought were better than Lyell's own proposals. One was a theory for the origin of coral reefs. Another was the recent elevation of the Cordillera. At a deeper level, too, he adopted Lyell's creed of gradual change. 'The science of geology is enormously indebted to Lyell – more so, as I believe, than to any other man who ever lived.'[12] Another tribute was paid in a private letter that Darwin wrote after the *Beagle* returned:

> I always feel as if my books came half out of Lyell's brain, and that I never acknowledge this sufficiently... the great merit of the *Principles* was that it altered the whole tone of one's mind, and therefore that, when seeing a thing never seen by Lyell, one yet saw it partially through his eyes.[13]

Without Lyell, it could be said, there might not have been any Darwin: no intellectual insights, no voyage of the *Beagle* as commonly understood. Darwin's thoughts began to circle around the notion of small changes leading to large effects. In doing so, he took one of the most important conceptual steps of his personal journey. For the rest of his life, he believed in the power of small and gradual changes. Afterwards, when working on evolution, he used the same concept of small and accumulative changes as the key to the origin of species.

At last the ship turned homewards and Darwin began to review his achievements. All the evidence points to the conclusion that he did not develop a theory of evolution on the voyage. Instead, he returned full of ideas and scientific ambition, determined to make sense of the riot of information he had acquired. Few young men ever had such an opportunity to see a world in its entirety. He was deeply impressed by nature's prodigality, the colour, variety and abundance on the one hand, and raw struggle and harshness on the other. And even though he came gradually to discredit the Bible as an authoritative record of real events, he was unwilling completely to give up his belief, partly because of this intense appreciation of nature's marvels. While standing in the middle of the grandeur of a Brazilian forest, he declared, 'it is not possible to give an adequate idea of the higher feelings of wonder, admiration, and devotion which fill the mind'.

He thought about the future too. For much of the voyage he apparently still intended to take up life as a country parson, although this prospect became progressively less appealing as his confidence as a naturalist grew. Towards the end, however, he told his sisters that he wished to pursue natural history as a vocation and hoped to be accepted as an equal in the scientific community. He wanted to be an independent gentleman-expert like Lyell, free to write books and follow his natural history inclinations, not tied to a university like Henslow, nor to the church patronage system like Fox. As the mental image of the parsonage among green English fields crumbled, there was the figure of Lyell beckoning behind it. 'It appears to me, the doing what little one can to encrease [sic] the general stock of knowledge is as respectable an object of life, as one can in any likelihood pursue.'[14] This shift in his ambitions rested on a conviction that he had new and noteworthy things to say. It also depended on the goodwill of his father in releasing his inheritance.

Darwin stepped on to the dock at Falmouth in October 1836 a changed man but not yet an evolutionist.

CHAPTER 2

'A theory by which to work'

Five years away on an Admiralty ship was a long time. When Darwin looked about him he could not help but notice how much England had changed. Railways were snaking across the land where stagecoaches had once travelled, towns crept relentlessly outwards, shops, chapels, factories and newly built churches sprouted everywhere. This was the England of Dickens's classic tales.

It is often hard to remember just how unstable British society was in these first four decades of the nineteenth century. The nation came as close to revolution as it ever had: conflict between landlords and manufacturers, workers against masters, province versus metropolis, the hungry and mutinous threatening the commercially minded, individualistic middle classes. Benjamin Disraeli's imagery of two nations, rich and poor, was not fanciful. 'The People's Charter', drawn up in 1838, with its famous six points – suffrage, the ballot, equal electoral districts, abolition of property qualifications, payment for MPs, and annual parliaments – frightened the political establishment deeply. A huge demonstration in 1839 ended in a bloody confrontation with the military. The last great Chartist rally on Kennington Common in 1848, although more peaceable and socialist in character, additionally reflected the agonies of Irish famine and political suppression. Karl Marx, surveying Britain through the eyes of his mill-owning friend Friedrich Engels in the 1840s, argued that capitalism was doomed to choke on its own surplus.

That this did not happen was substantially due to the dramatic industrial development that we think of as characteristically Victorian. From the 1850s a new and varied economy soaked up excess capital and diversified the labour force. Only three years after the Chartist demonstrations, people poured into London to visit the Great Exhibition of All Nations housed in the Crystal Palace designed by Joseph Paxton. Factories boomed. Steam technology and carboniferous investment generated unprecedented feats of engineering, and advances in transport systems brought the possibility of progress to nearly every corner of the nation. Railways were 'the ringing grooves of change' to Alfred Lord Tennnyson. By the time Darwin published *Origin of Species*, everywhere there was diversification, specialization and improvement.

This early fear of revolution is hard to recapture nowadays. There was widespread unease about any social or political activities that threatened the status quo. Prime among these were evolutionary notions: publicly to adopt transformist ideas was at that time to brand oneself as a dangerous political radical. Most notorious of all were the two men

Darwin had already read, Jean-Baptiste Lamarck (1744–1829) and his own grandfather Erasmus Darwin (1731–1802). Between the years 1798 and 1809, Lamarck and Erasmus Darwin had independently proposed that animals and plants were not directly controlled by a divine creator but spontaneously generated out of inorganic materials. Organisms then progressively advanced and diversified, they suggested, by adapting themselves to different environments. Both men believed that animals (and to some extent plants) became adapted by the use or disuse of various parts of their bodies and that these adaptations were passed on to the offspring – the inheritance of acquired characteristics, as it became known. They included humans in their schemes, and declared that humans would improve over time. Even the structure of society would be transformed. Other writers of the day played with the same idea, including the wit Thomas Love Peacock who made an orang-utan the gentlemanly hero of his comedy, *Headlong Hall* (1816).

Generally speaking, Erasmus Darwin's views were more accessible to the educated reading public than Lamarck's closely argued academic tomes because they were published in vigorous rhyming poems glorifying fecundity, human ingenuity and invention. Twenty years later, with the atheistic doctrines of French *philosophes* still ringing in British ears, and the revolution and Napoleonic wars fresh in the memory, such opinions were frequently associated with anti-theological activism, public protest by the workers and subversive exhortations to overthrow a useless British aristocracy.

Ideas like these circulated more or less underground in the medical circles that Charles Darwin briefly encountered at Edinburgh University. By contrast, Cambridge University, a training ground for the political establishment, took the lead in promoting an alternative: the clockwork universe of natural theology, the doctrine that Darwin had enjoyed as a young man but was ultimately to overthrow.

The world to which Darwin returned seethed with change and ideas about change. He felt that he was changing too. The years that he lived in London were the most intellectually creative that he ever experienced.

Naturally enough, he worked hard distributing his *Beagle* specimens to appropriate experts. Connections blossomed, publications were arranged. With Henslow's help, Darwin obtained a Treasury grant to publish formal descriptions by experts of his animal collection in *The Zoology of the Voyage of H.M.S. Beagle* (five parts, 1839–43). This handsome set of volumes was lavishly illustrated with hand-coloured plates and is one of the most attractive of Darwin's publications. He simultaneously composed a lively travel narrative drawing on the daily diary he had kept throughout the five years of the voyage. Published in 1839, under the title *Journal of Researches,* now usually called *The Voyage of the Beagle,* this brought him renown as an author. The great Alexander von Humboldt wrote to him to call it 'happily inspired', an 'admirable book... You have an excellent future ahead of you'. Such words from the man whom Darwin had idolized in his Cambridge days, and whose writings were generally regarded as the height of literary style, were praise indeed. The *Journal of Researches* remains the best-loved of all his writings.

Darwin also joined the Geological Society of London, where he delivered three short papers describing some of his geological results and met Charles Lyell for the first time. Lyell was overjoyed to find someone who was so appreciative of his *Principles of Geology* and the two became close friends. Everything about Lyell's personality chimed with Darwin's. 'I saw more of Lyell than of any other man both before and after my marriage… His delight in science was ardent, and he felt the keenest interest in the future progress of mankind. He was very kind-hearted and thoroughly liberal in his religious beliefs or rather disbeliefs; but he was a strong theist. His candour was highly remarkable.'[1] The two men spontaneously began to see each other nearly every day. Within a few months of his return, Darwin had achieved his ambition of joining the elite world of metropolitan science as an equal: he was elected to the Royal Society, the Athenaeum Club (the influential gentlemen's club in London) and the councils of the Geological Society of London and Royal Geographical Society. He 'went a little into society'.

All that was missing was a wife. Towards the end of 1838 Darwin felt established enough to propose to his cousin Emma Wedgwood whom he had known since he was a boy. She was the youngest daughter of the same uncle Wedgwood who smoothed his path on to the *Beagle* in the face of Dr Darwin's objections a warm-hearted, good-natured and supportive woman who married him for love and cared for him through thick and thin ever afterwards. They made a close and contented couple, united by affection and an interrelated gang of cousins, sisters, brothers, parents, aunts and uncles, and eventually a large brood of children of their own. They were married quietly at the Wedgwoods' home in Staffordshire in January 1839. Private income derived from mutual family investments enabled the couple first to live in London, and then, with the addition of a young family, in 1842 to purchase Down House and some twenty acres of land in the village of Downe, near Bromley in Kent. Darwin lived and worked at Down House for the rest of his days, a respectable and respected member of the village community.

This was not all. Early in 1837, some four or five months after arriving back in Britain, Darwin became convinced that species originated without divine agency. Oddly enough, for all the historical research into Darwin's breakthrough, it is not known precisely how or when this conviction came about. Of course, in some sense the genesis of any original insight is something of a mystery – all the greatest scientists have spoken of the unexpected way that a new thought or shift in perspective has dropped into his or her awareness. The words they use to describe this process, often akin to a revelation, or 'new eyes', seem to them inadequate for the effects that it precipitates. Mostly they agree that their minds were well prepared, often from years of thought, and that a multitude of factors brought them to that particular point, some personal, some intellectual, some circumstantial, some impossible to articulate, others deeply social and political. And of course historians have picked over Darwin's manuscripts to identify crucial passages in his *Beagle* notes and his early London publications that hint at the way his mind was going. In a small notebook written round about then, he struggled to put embryonic ideas into words. The origin of species ought to be just as comprehensible as the birth of individuals, he wrote: 'They die without they change, like Golden Pippins, it is a generation of species like generation of individuals... If species generate other species, the race is not utterly cut off.'[2]

A sense of uncertainty emerged primarily from the Galápagos birds. These birds were classified in March 1837 by John Gould, a taxonomist from the Zoological Society, who also helped Darwin with his large illustrated book, the *Zoology of the Beagle*. Gould identified several species of ground finch, with beaks differently adapted to eat insects, cactus or seeds, and put the mocking birds into three separate species. These species probably lived one to each islet, but Gould could not be sure because Darwin had not labelled them with their location. Surprised, Darwin mulled this information over. If each island had its own birds, as Gould suggested, his shipboard speculations about the instability of species were truer than he thought. Perhaps the similarities could be explained if the finches had diversified from a common ancestor?

He began recording a flurry of ideas in a series of private notebooks that he labelled A to E, and then M and N, now known as the Transmutation Notebooks. From the moment of opening Notebook B, around July 1837, he expressed the belief that some kind of evolution had taken place, not just among the Galápagos Island birds but involving everything, including humans. Entries in the notebooks tumbled over each other. Page after page, he built theories that stretched as far as his imagination would take him. Little of this would have occurred to the Cambridge undergraduate six or seven years before. The *Beagle* expedition naturally was the baseline for many of his speculations. Yet he also revisited Dr Erasmus Darwin's theories, and pondered Lamarck's writings. All the time he read voraciously and questioned knowledgeable contemporaries. Even before he devised the theory that ultimately appeared in *Origin of Species,* he saw important parallels between domestic breeds and wild species. That analogy was to remain at the heart of his work.

From the very first, he regarded human beings as members of the animal kingdom and hoped to explain our origins without any reference to God's creation, a theme that took him far into 'metaphysics on morals', as he called it. 'Man from monkeys?' he asked himself. 'Man in his arrogance thinks himself a great work, worthy the interposition of a deity. More humble and I believe true to consider him created from animals.'[3] Some of these 'mental riotings', as he dubbed them, took him very far along the road of materialism, the philosophical doctrine of believing that there were no spiritual or divine forces in nature, only matter. If he denied the createdness of everything, where did that leave human beings and our hopes of salvation? Our thoughts are merely secretions of the brain, he alleged. 'Oh you materialist!' he exclaimed, half in admiration at his boldness.

All along he sought an explanation for the way that animals and plants might actually change. This was dramatically brought into focus after he read in September 1838 *An Essay on the Principle of Population* (1798) by the British economist Thomas Robert Malthus.

Malthus's intention was to explain how human populations remain in balance with the means to feed them – his essay was an important contribution to the social and political economy of Britain in the 1790s, presented as a rational examination of the natural laws of society. By the 1830s, his impact on British life had deepened far beyond what even he may have hoped, for Malthusian doctrines had come to dominate government policy. The argument was starkly simple. The natural tendency of mankind, Malthus said, was always to increase. Food production could not keep up. Yet there was an approximate balance, he claimed, because the number of individuals is kept in check by natural limitations such as death by famine and disease, or human actions such as war, sexual abstention and sinful practices such as infanticide. These checks, he claimed, were a necessary part of human existence. Malthus went on to say that such checks usually fell on the weakest – the poorest and sickest – members of society. It was God's will that this should happen. One consequence, Malthus warned, was that giving charity to the poor would simply encourage more reproduction and greater food shortages. In decades to come, these opinions were reflected in food riots, controversy over the Poor Laws and public reaction against the Corn Laws. The passing of the Poor Law Amendment Bill in 1834 introduced the Victorian response to this social and economic issue in the shape of the workhouse, where local parish charity was replaced with people having to work for their bread.

Darwin lived in this world too. He moved in broadly the same circles as Malthus, and was acquainted with some of the people who knew Malthus before his death in 1834, including Fanny Wedgwood (Darwin's sister-in-law) and the author Harriet Martineau who wrote Malthusian tracts for the respectable classes. It is only to be expected that in the midst of this topical political concern over Malthusian issues that Darwin picked up a copy of the original book and settled down to read it.

The moment was recorded in Notebook D in an entry dated 28 September 1838. Too many individuals are born, he paraphrased from Malthus. There is a war in nature, a struggle for existence. In the fight to live, the worst or weakest organisms tend to die first, leaving the better forms, the healthiest or better adapted. These survivors would be the ones that generally had offspring. If such actions were repeated over and over again, organisms tend to become ever more appropriately suited to their conditions of existence. He called the process 'natural selection', meaning a process in the natural world analogous to the 'artificial' selection that he saw farmers and horticulturists applying to domestic animals and plants. Farmers disposed of the worst and saved the best for breeding purposes in order to create a faster greyhound or a woollier sheep. In the wild, Darwin suggested, it was nature itself that did the selecting. In short, he hit upon a way of explaining Paley's perfectly designed adaptations without any reference to a creator. 'Being well prepared to appreciate the struggle for existence... it at once struck me that under these circumstances favourable variations would tend to be preserved, and unfavourable ones to be destroyed. Here, then, I had at last got a theory by which to work.'[4]

This was the essence of Darwin's theory, scarcely to change except in one major point until it was published twenty years later in *Origin of Species*. He recognized its explanatory power, saying it would revolutionize the biological sciences. He acknowledged its religious implications, not only for the new vistas it opened up for the possible origin of mankind but also for the way it would deny any role for God in nature and challenge the natural theological traditions so firmly embedded in British life and institutions. These were hazardous views. They were the antithesis to the harmonious world of perfect adaptations preached by Darwin's old friends and teachers like Henslow and Sedgwick. Even brave, forward-looking Lyell might object.

For the moment, Darwin kept this theory secret. He realized the need to be cautious. It may have struck him as too sudden, too dangerous and unorthodox, too much in need of fuller and further reflection. He saw no need to rush into print. But he told Lyell that he was filling 'notebook after note-book... with facts which begin to group themselves clearly under sub-laws'. Displaying extraordinary self-discipline he worked intensively on these ideas in private.

Only his wife Emma was aware of his general notions. She knew Darwin had religious doubts. Even before they married she said. 'My reason tells me that honest and conscientious doubts cannot be a sin, but I feel it would be a painful void between us.'[5] She expressed the fear that science was leading him into ever-greater scepticism. Hesitantly, she suggested that Darwin's doubts might prevent them meeting in the afterlife or belonging to each other forever. This letter was treasured by its recipient. 'When I am dead, know that I have many times kissed and cryed over this', Darwin wrote on the edge.[6] There is every indication that by the time he was writing about natural selection in his notebooks – the same year that he married Emma – Darwin had dispensed with most formal religious structures while still believing in some supernatural force beyond human knowledge. He was not an atheist, however. In fact, it seems that he was never an atheist, not even at the height of the controversy that followed publication of *Origin of Species*. He said in his *Autobiography* that he thought about religion a good deal during these years, and that the term 'theist' was probably closest to what he felt. Later he called himself agnostic, a word coined by his friend Thomas Henry Huxley. Nor did Emma have anything to fear about his behaviour or sense of morality. He was basically a good man, humble and kind, and always did his best to act according to the traditional values that he had learned as a child.

All this came at a terrible price. Slowly Darwin sank into chronic ill-health. From the time of his marriage, he increasingly felt nauseous and began to suffer from recurrent headaches; occasionally he experienced bouts of actual vomiting, sometimes stretching over several weeks. There were long periods of unaccustomed weakness and debility. Whether this was directly related to his evolutionary ideas is hard to say. Such is usually assumed to be the case. To it might be added his religious qualms, a self-imposed punishing work schedule, the ceaseless publishing activities, duties in London's learned societies and worries about the future. There is little evidence for physical causes for his ill-health such as arsenic poisoning, allergies, lupus or Chagas disease, a South American disorder transmitted by the black bugs of the pampas that Darwin may have picked up during his travels. Some 150 years later it is probably futile to try to diagnose his condition. For the rest of his life, apart from one or two brief interludes, ill-health was integral to his personality, work and lifestyle.

Darwin tried the water cure for many years, where he followed a regime of wet-sheet packing, rubs, enemas and douches, mostly attending the famous Dr James Gully in Malvern but also visiting centres in Surrey and Yorkshire. Later he took professional advice for dyspepsia. Modern-day psychiatrists have remarked on the self-attentive, tightly-packaged personality that these medical encounters suggest.

One curious consequence can be found in his 'health diary', a daily record of how he felt, that he kept for three or four years, marked up with symbols to indicate whether he felt ill, 'very ill', or merely 'poorly'. Perhaps after the robust good health of the *Beagle* voyage, Darwin could never truly let go. By the time of his death, the whole family were almost professional invalids, plagued by weak pulses, nausea, chronic debilities, headaches and undefined stomach troubles. It seems unfair to lay these disorders solely at the door of a hypochondriacal parent, but this does seem likely.

This disorder, or constellation of disorders, has naturally intrigued historians. All seem to agree that Darwin's ill-health in some way must have reflected pressures generated by the subversive theories he was developing in private. Scholars with psychoanalytic leanings tend to explore the motif of a creative malady or the 'madness' of genius. They call on the notion built into literary and artistic studies that works of great originality are usually produced in a state of intense mental turmoil – that creativity emerges from extreme emotion, often at the edge of sanity, or brings about the physical wreckage of the human frame as it gives birth to an artistic masterpiece. Yet it should also be acknowledged that Darwin's letters during those years of thinking and writing never expressed any outright mental torment. Perhaps Freudians would propose that these feelings were sublimated. Darwin certainly experienced the fear of rejection mingled with a high anxiety that his life's work might be damned or ridiculed, and that his evolutionary theory was, in effect, murdering the God of the ancients. If such feelings were tightly controlled in the Victorian domestic context then it seems entirely possible that Darwin should find the only possible way to express his alarm was through undiagnosed, sub-clinical disorders. Alternatively, a growing number of historians argue that Darwin suffered from some real bodily ailment and try to match modern illnesses to the symptoms he described so piteously in letters and diaries. Few agree on what it might be. There are consequently marked factions in scholarly circles, one group arguing for psychological reasons for Darwin's illness, the other for physical causes. At root, they diverge over the definition of creativity and the role ill-health might play in inspiration and imagination.

As time went by, Darwin cautiously described some of his evolutionary views to close scientific friends, gauging their reaction. In June 1842 he felt he had the shape of the theory sufficiently well formulated to write a short private sketch, which he expanded into a longer essay in 1844. One notable thing about these manuscripts was the absence of any reference to the origin of mankind and the creator. Possibly his talks with Emma about religion had motivated him to avoid discussion of human beings. Or he might have decided that he needed to know a great deal more before he could argue convincingly about mankind's origins. Whatever the reasons, he systematically drained the manuscript of human beings, ensuring they did not return until long after *Origin of Species* was published.

This essay of 1844 could easily have been published as it stood. Indeed, that was partly Darwin's intention. He entrusted it to his wife with a letter to be opened in the event of his sudden death, stating that she should engage an editor to publish it posthumously. 'If it be accepted even by one competent judge, it will be a considerable step in science,' he said.[7] Puzzlingly, however, he then put it to one side. It was never published in his lifetime. Was this a deliberate delay? Was he scared of publishing? Many people think so. And yet, in a sense, he was not in any hurry. Nowadays, in the light of all that is known about his personality and correspondence, it seems feasible to suggest that a strong commitment to scientific accuracy and a proper sense of scientific caution were at least as high in his mind as any fear of the consequences of publication. He did not feel ready. The scale and scope of his musings in the Transmutation Notebooks indicate the very wide range of investigations and topics that he thought were relevant. He had barely started to chip away at the surface of them.

One striking event, however, gave him reason to pause. This was the publication in 1844 of an anonymous evolutionary book, *Vestiges of the Natural History of Creation*. This book dramatically changed the texture of debate on evolution – firing up the theologians, pushing secular thought uncomfortably into Victorian drawing rooms, inspiring violent criticism on the one hand, and fascinated attention on the other. *Vestiges* became a popular publishing phenomenon on a scale similar to Tom Paine's *Rights of Man*. It raced across the

English-speaking world in cheap editions and made a splash in other countries in translation. The unknown author wrote fluently of the self-generated development of the living world from specks of animate matter to men and women. Although the scientific content was on the whole weak, and the proposed mechanisms of change at times laughable, its general evolutionary thrust was clear. It was a book that tapped into the progressive aspirations of the age. One satirical cartoon published in Melbourne, Australia, caught the nub of the issue by showing the local Chinese workforce transmuting into Western gentlemen.

Whether intrigued or disturbed, people discussed *Vestiges* intently in journals, parlours and meeting halls. Venomous scorn poured from Adam Sedgwick, Darwin's old geology professor. Sedgwick accused 'Mr Vestiges' of making philosophy out of moonshine. The book was so uninformed, so inaccurate, so contentious and unsupported by proven facts, he ranted, that it could have been written by a woman. The real issue at stake, continued Sedgwick, was the origin of human beings and the moral status of mankind. *Vestiges* ignored the Garden of Eden, the creation of Adam and Eve, the expulsion from the garden and eventual covenant with God, and suggested that we came from orang-utans.

In fact, the author was not a woman, but Robert Chambers, a successful Scottish journalist who founded, with his brother, *Chambers's Edinburgh Journal*, a weekly magazine containing numerous short articles on literature, science and industry, manners and morals, all interspersed with poems and stories. Robert Chambers was an enthusiastic proponent of self-education and the doctrine of phrenology, in which it was believed that an individual's character could be read from the shape of the skull and that various faculties of the mind could be enhanced or reduced by willpower and training. Beyond this Chambers was deeply interested in the science of the day. His book deliberately discussed the things avoided by more conventional scientists, for example the possibility of insects being created by electricity. It was published anonymously because he knew the storm that would follow. When people afterwards said that evolution was in the air, *Vestiges* was what they meant.

Darwin was stunned. Reading *Vestiges* in November 1844, so soon after completing his own essay on evolution, was a shock from which he took years to recover. In very broad terms, the general thesis in *Vestiges* was startlingly similar. Admittedly, Chambers set his proposals within a vast panorama of the cosmos, which Darwin did not plan to do. Admittedly, he included mankind, which Darwin studiously avoided. And many of *Vestiges*'s facts were incorrect. But the book grasped the essential principle of gradual, natural origins. *Vestiges* formed the closest to a rival that Darwin had as yet met. He was forced to acknowledge its influence, take account of its arguments, and show where he differed. Yet he must also have winced at the abuse raining down on the anonymous author. Was he in for the same treatment? One new friend, the botanist Joseph Hooker, chattered away in letters, saying that he was delighted by the multiplicity of themes that *Vestiges* brought together. Darwin was stung. 'I have also read the *Vestiges*, but have been somewhat less amused at it, than you appear to have been,' he said stiffly. 'The writing and arrangement are certainly admirable, but his geology strikes me as bad, and his zoology far worse.'[8] Obsessively, he began to build up his own edifice of dependable factual information that would be so much admired when he eventually published *Origin of Species*, and which lifted his book far above the ordinary.

Over the next fifteen years or so, Darwin worked relentlessly to find support for his theory. Energetically he tackled a programme of experiments in his garden, especially taking up the hobby of breeding pigeons with the advice of a noted fancier, William Tegetmeier. Pigeons were his 'current love' he told Hooker enthusiastically, and a hit with the young family. In them Darwin sought direct observational evidence of the inheritance of traits such as black wing feathers or the reversion of fancy breeds back to an ancestral type. In a manner of speaking he tried the same thing with plants in the greenhouse, but here he was looking for evidence of variability and how incipient species became mutually sterile. Many of these experimental questions were raised by him in brief notices and articles in popular natural history magazines. There was never a final conclusive experiment, evolution was not that sort of theory. Instead, Darwin continued to research and formulate new questions, consult printed literature and correspond across the globe with a vigour that would, with hindsight, amaze him when he came to write his *Autobiography*.

As part of this extensive programme, Darwin studied barnacles. Nearly a barnacle himself by now, hunched in his study writing letters or compulsively making experiments in the garden, he hardly liked to leave home at all. He threw himself into a systematic inquiry into every known species of barnacle, living and fossil, an unusual initiative on his part that took him eight years to complete. Historians tend to smile at so much time spent on insignificant organisms and call it a sideline, a delaying tactic in order that Darwin might avoid confronting the furore that would arise out of publishing his other more wide-ranging evolutionary views.

So it may have been. What he found in barnacles, however, brought important shifts in his biological understanding, strengthened his belief in evolution and provided an essential backdrop to *Origin of Species*. Every day, he looked at barnacle structures as the result of evolution. He searched for the tiny adaptations that made one form more successful than another, for the diversification that led to increasingly specialized forms, saw how one organ (such as the ovarian tract) could become adapted to perform an entirely different function (such as the cement gland) and scrutinized the unusual reproductive strategies of the group. Most important of all, this study of barnacles revealed the high rate of variation that occurred in nature. Yet Darwin never mentioned evolution in the barnacle monographs that he published in 1852 and 1854. It is only now that we can see just how heavily his assessments depended on concepts that he wished to keep private.

The Royal Society of London was impressed. This premier society for science awarded Darwin its Royal Medal in 1853 for the barnacle books in conjunction with his publications on the geology of South America. Afterwards he was amused to be lampooned by the novelist Edward Bulwer Lytton as an absurdly preoccupied 'Professor Long'. The work had taken so many years, indeed, that one of Darwin's children believed that all fathers spent their days looking at barnacles through a microscope. 'Where does your father do his barnacles?' Leonard Darwin innocently asked a young friend.

Tragedy struck in the midst of this quiet domestic activity. No Victorian was ever immune but death in the family hit Darwin and his wife deeply. Their second child, the daughter they called Annie, died in 1851 of an unidentified fever, aged ten years old. She was the apple of Darwin's eye. By then the Darwins had had eight children: William (b.1839), Anne (b.1841), Henrietta (b.1843), George (b.1845), Elizabeth (b.1847), Francis (b.1848) and Leonard (b.1850). One little girl had died at only three weeks old in 1842; two additional sons would be born later, Horace in 1851 and Charles in 1856. Charles, the last, died from scarlet fever, aged about eighteen months. Annie's death may have tipped Darwin finally into disbelief. The doctrines of the Bible in which Emma took comfort were hurdles that he could not jump. In a short memoir that he wrote for his and Emma's eyes only, in which he praised Annie's sunny nature, the despair can easily be read. How could a caring, beneficent creator extinguish such an innocent child? How could God make a child suffer so? His science told him that Annie was irretrievably gone. After this, Darwin turned back to his work with a new grimness – an edge of determination that helped him carry on where other men might have abandoned their studies.

Even the keenest grief could not last forever, and Emma produced the next baby and other diversions eased into view, including the Great Exhibition in Joseph Paxton's glittering Crystal Palace later in the year. During this period of study Darwin devised what he called 'the principle of divergence', the only major alteration that he made to the original theory of natural selection that he had formulated some twelve years before. It was an important innovation, for he needed it to explain how natural selection could produce the branches of the tree of life. With a leap of the imagination that might have been stimulated by the family visit to the Great Exhibition, he drew on industrialized England for a metaphor. Natural selection probably favoured those animals and plants that diversify, just as if nature were a factory bench at which production was more efficient if workers performed different tasks. This was an industrial analogy ubiquitous in the decades of specializing workforces, personally familiar to Darwin from the Wedgwood china company. The most successful variant, he said, was the one that could seize on unexploited places or roles in the natural economy. 'I overlooked one problem of great importance… and I can remember the very spot in the road, whilst in my carriage, when to my joy the solution occurred to me; and this was long after I had come to Down.'[9]

These adjustments, and the long-term concentrated attention that Darwin gave to his idea, created the circumstances in which his theory could flourish. In his mind it was now robust. It could withstand publication. The long years of private reflection, he came to think, had fulfilled their objective. When a group of friends came for a weekend visit to Down House in April 1856, he felt ready to argue his case with them. All fairly critical of the biology of their day, these friends discussed the failings of current definitions of species with Darwin. They cheekily 'ran a tilt at species' said Lyell, hearing about the weekend afterwards. And when Lyell visited a few days later, he too fell into a deep discussion with Darwin about transmutation that he recorded in his diary. Lyell warmly encouraged Darwin to publish. Evolution was in the air, he said. He drew Darwin's attention, not to *Vestiges*, although this was probably mentioned, but to an article in a popular natural history magazine by a relatively unknown naturalist called Alfred Russel Wallace (1822–1913). Wallace discussed the relationships between varieties and species and – if one had the eyes to see – implied a real continuity between them. It was time for Darwin to publish, said Lyell. There were other men, other theories.

The warning took root. 'Began by Lyell's advice writing Species Sketch,' Darwin recorded solemnly in his diary on 14 May 1856. 'I am like Croesus overwhelmed with my riches in facts,' he went on to tell his cousin Fox, now a clergyman, 'and I mean to make my Book as perfect as ever I can.'[10]

CHAPTER 3

Publication

Despite Darwin's measured calm, his book was actually born in crisis. The story has often been told. For more than two years he carefully composed a long manuscript, a big book that he planned to call 'Natural Selection'. Few of his friends knew what he was doing, although his web of correspondents circled the globe, feeding his insatiable appetite for facts. Picking up a thin well-wrapped package one morning in June 1858, he wondered who could be writing to him from Ternate, an island in the Dutch East Indies halfway between Celebes and New Guinea. He hoped it might contain some news about exotic species. However, here in a short handwritten essay, the naturalist Alfred Russel Wallace set out his own account of evolution by natural selection. The date that the essay arrived will never be known for sure. But late in the evening of 18 June 1858 Darwin wrote to Lyell to express his despair at being well and truly forestalled. 'I never saw a more striking coincidence... if Wallace had my MS sketch written out in 1842 he could not have made a better short abstract!'

Deeply surprised that someone else had come up with the

same theory, he consulted his two closest friends, Lyell and Hooker, about what he should do next. Scientific convention and gentlemanly honour indicated that he should bow out and let Wallace take the credit. None the less Lyell and Hooker felt that Darwin should not lose his claim to be the originator of the theory. They were aware of the lengthy manuscript that Darwin was working on. There was room for manoeuvre, they insisted. They therefore proposed that they should send Wallace's essay forward for publication along with a short account of Darwin's own findings. There would be a double announcement and priority would be shared. Doubtfully, Darwin agreed. 'I cannot tell whether to publish now would not be base & paltry; this was my first impression, and I shd. have certainly acted on it, had it not been for your letter.'[1]

This double announcement took place as suggested on 1 July 1858 at a meeting of the Linnean Society of London, the leading scientific society for natural history in Great Britain. As it happened, Lyell and Hooker were influential in the Society's administration. They managed to rush the double paper on to the programme of an unexpected extra meeting that was taking place at the end of the season, rescheduled because of the death of the botanist and former president of the society, Robert Brown.

Oddly enough, considering the content, there was little excitement among the small audience when the papers were read out loud by the secretary, although when they were published in the Society's journal a few months afterwards several people recognized their likely impact. Neither Darwin nor Wallace was at the Linnean Society meeting itself. Darwin's tenth child, still just a baby, was dangerously ill with scarlet fever and died on 28 June 1858, only two days before the announcement. As a loving father Darwin felt too wretched with grief to attend. Wallace was miles away in the Far East. Indeed, he knew nothing about it. With postal services to the opposite side of the globe taking three or four months, he had yet to receive the letter that told him that his essay duplicated another man's work and was being made public in a twosome. When he did find out, he admitted that he was astonished. Courteous and mild by nature, he immediately wrote to Darwin and the others to say that he thought the publication arrangements were entirely satisfactory. Even though Darwin is usually characterized by biographers as generous and gentlemanly during this incident, the real generosity surely rests with Wallace, the unwitting catalyst for the commotion. Historians have often wondered subsequently if Wallace was short-changed, or even exploited, by the arrangements made by Lyell and Hooker and agreed by Darwin.

For there is no disguising the fact that Wallace came from the other end of the Victorian social scale. Self-educated, and with no private income, he made an unsteady living by collecting natural history specimens to sell to museums and collectors. His first collecting trip had been to Brazil with his friend, the naturalist Henry Walter Bates, to comb the Amazonian rainforest for rare birds and insects. Then, in

1853, he struck out independently to the Malay Archipelago, where he stayed for eight years, travelling some 14,000 miles within the region. It was the possibility of acquiring some Malaysian fowl that originally brought him into Darwin's network, and they had occasionally corresponded about specimens. When Lyell had drawn Darwin's attention to Wallace's earlier article in April 1856, Darwin wrote to Wallace to praise it and, in passing, mentioned his current work on definitions of species and varieties, a topic of great practical interest to naturalists at the time. It was probably this polite expression of interest that encouraged Wallace to send his evolutionary essay to Darwin in 1858.

Wallace's personal circumstances and aspirations were very different from Darwin's. Nevertheless, he read many of the same books, encountered many of the same biological problems during his overseas travels and shared much the same forward-looking Victorian milieu. Inspired by *Vestiges'* world of constant progressive development, he eagerly adopted the concept of transmutation. He hoped to find in Sumatra or Borneo evidence that humanity had formerly emerged from the great apes of the region. Fine observational skills had already led him to match the geographical distribution of butterflies in the Amazon river basin with their variation, an observation that served the same function in his intellectual development as Darwin's Galápagos finches. He read Lyell and saw, like Darwin, that gradual geological change might indicate equally gradual changes in species. He read Darwin's account of the *Beagle* voyage. He read

Malthus, and took from him the same notion of differential survival. Wallace even had a 'Malthusian moment' akin to Darwin's flash of inspiration. Suffering from malarial fever, Wallace was resting one day from an attack of the 'chills', pondering the human demography of the islands around Papua New Guinea, when he suddenly realized that the Papuan population was being gradually exterminated by invasions of Malays. As for Darwin, so for Wallace: everything fell into place. He wrote, using the same vocabulary as Darwin, of a 'war' in nature, competition between individuals and the triumph of the more successful form.

Sitting alone in his study, always working, always concentrating on the job in hand, Darwin had allowed himself to feel that he was in no danger of being pre-empted, no need to hurry, until the letter came from Wallace. Yet as Lyell hinted, there were plenty of proponents for advanced schemes of thought if one had eyes to see.

Great currents of change were making their presence felt in Britain. High-level critical thought about the Bible was spreading as the biblical scholars of Europe investigated sacred texts as if they were solely historical documents. Inside the secluded quadrangles of Oxford colleges, the Reverend Baden Powell frankly discounted miracles, while John Henry Newman converted to Catholicism and initiated the Tractarian movement. George Eliot's translation of Strauss's *Life of Jesus* (1846) presented the Son of God to English readers as if he were an ordinary man. One by one, Victorian thinkers claimed the right to investigate the world around them without recourse to either God's miraculous powers, or the Bible's word, or the church's doctrinal authority. Some, like Tennyson or Matthew Arnold, began seriously to doubt the religious system in which they had been raised. In the elite world of British literature and letters this movement ultimately manifested itself in the book entitled *Essays and Reviews* (1860), in which seven eminent theologians challenged traditional interpretations of scripture. Anxious doubts, secular inclinations and dissatisfaction with conventional doctrines were launched among intellectuals long before Darwin came on the scene.

The men and women of the influential liberal magazine, the *Westminster Review*, led by the charismatic editor John Chapman, and Mary Ann Evans (the novelist George Eliot), for example, were fascinated by the idea of inbuilt natural laws and steady advance in human society. Their friends, the historian Henry Buckle (1821–62) and philosopher Herbert Spencer (1820–1903), extolled development in society and nature. Buckle considered the history of nations, telling his readers that civilized societies will always overcome the less developed. From the barbarism of Ancient Rome to Victorian parliamentary democracy, Buckle's history books argued for progressive improvement. In Spencer's writings, the same ideas took the form of a law of development that he applied to animals and plants as readily as to politics, economics, technology and human society. In 1852 he published 'The development hypothesis' in which he supported a general Lamarckian theory of animal transmutation, followed by a

Malthusian-style essay on the 'Theory of Population' in the *Westminster Review*, where he wrote about population pressure driving the weakest to the wall. His anti-theological *Principles of Psychology* (1855) followed shortly afterwards, and by the end of the decade he had begun an ambitious, lifelong re-evaluation of metaphysics, the first part of which was published in 1862. Spencer believed that biological and social progress formed one broad evolutionary continuum – that they were governed by the same immutable laws and controlled by the same forces of nature. Darwin had never taken any of his writings seriously.

Spencer was not the only one to think like this. George Henry Lewes, the editor of the forward-looking *Leader*, regular contributor to the *Westminster*, and George Eliot's lifelong partner, delved into anatomy and physiology, proposing that human thought was merely a by-product of the brain's physiological activity rather than a gift from God. Supported by William Benjamin Carpenter, another physiologist, Lewes pushed divine agencies right to the background. Harriet Martineau shocked pious readers by declaring her religious doubts. Charles Kingsley, the author and radical clergyman, brought his social-realist novel *Alton Locke* (1850) to a climax with the hero's nightmarish dream of a metamorphosis from jellyfish to man. These lively modernist thinkers rejected natural theology, the system of explanation entrenched in the old universities, and opted for something more flexible and personal, a god who reigned unobtrusively in the background, who did not need the rigmarole of church doctrine.

By 1850 or so transmutation seemed less threatening to forward-looking thinkers such as these. Freshly sanitized by the mid-Victorian gloss of industry and commercial confidence, the dangerous, volatile air of the 1830s and 1840s lifted. Prosperity and progress appeared as motifs of the age. Middle-class liberals advocated self-improvement, literacy and education on the one hand, and public lectures and museums on the other. Medical men wondered about the possibility of the spontaneous generation of the smallest cellular beings and discussed Louis Pasteur's experiments with interest. Darwin's old acquaintance Robert Grant moved to London to become professor of zoology at University College London and lectured on the evolutionary scale of nature, until too old to continue. A fair number of leading intellectuals embraced doctrines of self-advancement, economic progress and the onward thrust of civilization to a greater or lesser degree, without necessarily overstepping the divide between faith and disbelief, and unknown numbers of less public figures, such as Alfred Russel Wallace, were contemplating the world through newly secular eyes. Obvious technological advances and economic expansion reinforced the point. Samuel Smiles's book *Self-Help* – the bible of the improving middle classes – highlighted the belief in entrepreneurial improvement sweeping into every arena of mid-century existence.

Darwin's unexpected collision with Wallace had one immediate effect. He was catapulted into writing *Origin of Species*. Immediately after the double Linnean Society paper had been read, he took his wife and family away for a brief holiday to recover from the death of baby Charles. Then, in the space of thirteen months, he produced a compact, tightly argued book.

In essence, Darwin drastically compressed the long manuscript he had already written. Afterwards he regretted losing so much of the solid scientific evidence he had struggled to collect and always regarded *Origin* as an enforced 'abstract'. For many years afterwards he still planned to publish the original long manuscript that had been interrupted by Wallace's letter.

Darwin called this shorter book 'one long argument'. And what an argument it was. Few scientific texts have been so closely woven, so packed with factual information and studded with richly inventive metaphor. Darwin's literary technique has long been noted for its resemblance to *Great Expectations* or *Middlemarch* in the complexity of its interlacing themes and his ability to handle so many continuous threads at the same time. Hardly daring to hope that he might initiate a transformation in scientific thought, he nevertheless rose magnificently to the occasion. His voice was in turn dazzling, persuasive, friendly, humble and dark. His imagination soared beyond the confines of his house and garden, beyond his debilitating illnesses and the fragile health of his children. At his most determined, he questioned everything his contemporaries believed about living nature, calling forth a picture of origins completely shorn of the Garden of Eden and dispensing with the image of a heavenly clockmaker patiently constructing living beings to occupy the earth below. He abandoned what John Herschel devoutly called the 'mystery of mysteries' and replaced Paley's vision of perfect adaptation with imperfection and chance. Animals and plants should not be regarded as the product of a special design or special creation. 'I am fully convinced that species are not immutable,' he stated in the opening pages.

Further than this, Darwin's underlying theme was gradualism. Everything happened little by little, just as Lyell claimed. Everything was linked by one and the same explanation. Time, chance and reproduction ruled the earth. Struggle, too. Those who sought a radically new manifesto for the living world were sure to find it in his words: no one could afterwards regard organic beings and their natural setting with anything like the same eyes as before; nor could anyone fail to notice the way that Darwin's biology mirrored the British nation in all its competitive, entrepreneurial, factory spirit; or that his appeal to natural law unmistakably contributed to the general push towards secularization and supported contemporary claims of science to understand the world in its own terms.

Another kind of narrative emerged as well, often mentioned by reviewers. Darwin wrote as he always wrote, in the same likeable, autobiographical style he had developed during the *Beagle* voyage and brought alive in his *Journal of Researches*. Much later on, his son Francis Darwin said this pleasant style of writing was characteristic of his father in 'its simplicity, bordering on naiveté, and in its absence of pretence... His courteous and conciliatory tone towards his reader is remarkable, and it must be partly this quality which revealed his personal sweetness of character to so many who had never seen him.'[2] Although his theories might frighten, his style was thoroughly sympathetic and genial, creating a distinctive magic between author and reader. He appeared in his book just as he appeared in life: as a reputable scientific gentleman, courteous, trustworthy and friendly, a man who did not speak lightly of the momentous questions coming under his gaze, a champion of common sense, honest to his data, and scornful of 'mere conjecture'. This humane style of writing was one of his greatest gifts, immensely appealing to British readers who saw in it all the best qualities of their ancient literary tradition and contemporary Victorian values. It served him well during the controversial years to come, defusing personal animosity and allowing even the harshest of critics at least to acknowledge his sincerity and meticulous investigation.

As an argument, *Origin of Species* was divided into two unequal halves. The first, shorter half set out the apparent facts of nature and led up to Darwin's presentation of the theory of natural selection in Chapter Four. The remainder of the book showed how the theory could explain or illuminate key biological areas such as embryology, classification, palaeontology and geographical distribution. An evocative conclusion invited readers to consider his point of view without prejudice. Unusually for a scientific book, Darwin also provided a frank discussion of the many stumbling blocks that would probably occur to readers, in a chapter called 'Difficulties on the Theory'. He admitted, 'Some of them are so grave that to this day I can never reflect on them without being staggered... I have felt the difficulty far too keenly to be surprised at any degree of hesitation in extending the principle of natural selection to such startling lengths.'[3]

This structure was carefully thought out. Natural selection is not self-evident in nature nor is it the kind of theory in which one can say 'look here and see'. Darwin had no crucial experiment that conclusively demonstrated evolution in action. He had no mathematical equations to establish his case. All these were to come a century later. Everything in his book required the reader's imagination. Like Lyell in his *Principles of Geology*, he had to rely on an analogy between what was known and what was not known. He depended on probabilities. He used words of persuasion, invited revisualization. Instance after instance was said to be 'quite inexplicable on the theory of independent acts of creation'.

The sheer variability of organisms came first. Every pig or cow, every blade of corn, as he described it, was in some way slightly variable. No two animals or plants were exactly alike. Farmers and horticulturists made use of these slight variations in individuals to improve a huge range of cultivated stock. Most of his readers would have agreed. The vast agricultural and horticultural wealth of the nation was based on exactly these activities, and large numbers of ordinary men and women possessed direct experience of the commonplace household animals and plants he described: dogs, gooseberries, cattle, garden flowers. 'Breeders habitually speak of an animal's organization as something quite plastic, which they can model almost as they please,' he said, and quoted Sir John Sebright, who claimed with respect to pigeons that 'he would produce any given feather in three years, but it would take him six years to obtain head and beak'.[4]

The biggest problem here, and one on which future critics alighted, was that Darwin had no knowledge of how the variations arose. He wrote *Origin of Species* long before the modern science of genetics was developed. The only thing that he could do was to demonstrate that variations indisputably did occur in domestic organisms. So his early pages were crammed full of examples drawn from every branch of natural history – a factual overkill that even reviewers noted. To this he added a matching account of variability in wild animals and plants. All his notes about barnacles' innards, donkey's stripes, primroses and oxlips took their place. Privately, he characterized this as a 'short & dry chapter'.[5]

The real point came next. Too many offspring were born. The living world teemed with deadly competition and slaughter, the same elemental energies, red in tooth and claw, that Tennyson characterized in *In Memoriam*. 'What war between insect and insect, between insects, snails, and other animals with birds and beasts of prey – all striving to increase, and all feeding on each other or on the trees or their seeds and seedlings, or on the other plants which first clothed the ground and thus checked the growth of the trees', wrote Darwin.[6] God's harmony was an illusion. Unsure whether he would be believed, he produced another flood of examples. Limited resources, limited places in nature, and continued natural fecundity gave rise to a battle for survival.

This was the point where he proposed the theory of natural selection. Harking back to the earliest and most powerful metaphor he had explored in his transmutation notebooks of the 1830s, Darwin declared that there was an important analogy between what happened in the farmyard and garden and in the natural world. In the same way as mankind can mould and adjust domesticated species to suit passing needs or tastes, so nature can pick the best adapted. The ones 'selected' to survive would be the parents of the next generation.

> It may be said that natural selection is daily and hourly scrutinising, throughout the world, every variation, even the slightest; rejecting that which is bad, preserving and adding up all that is good; silently and insensibly working, whenever and wherever opportunity offers, at the improvement of each organic being in relation to its organic and inorganic conditions of life.[7]

Ruminatively, Darwin elsewhere acknowledged the problems that this anthropomorphic language would generate. He often personified natural selection in *Origin of Species*. While this was perhaps unavoidable in the general sense, he frequently gave the impression that natural selection was an active agent. To some, it might even be thought of as God, a divine gardener in the sky, as it were, who chose the variants that were to succeed. Years afterwards Darwin admitted that this was not his intention and that he ought to have used a more neutral expression like 'natural preservation'. The same entanglement occurred when he used the word 'adaptation', which hinted at some form of purposeful strategy in animals and plants, the exact opposite of what he meant. Later, he used 'contrivance' as a partial solution. Over and over, Darwin struggled with his vocabulary. The language he had to hand was the language of Milton and Shakespeare, steeped in teleology and purpose, not the objective, value-free terminology sought by science.

He was not even able to speak of 'evolution' as such, because at that time the term was mostly used to describe the unfolding of hidden embryological structures; it was the ensuing debate around his published work that gave the word its modern meaning. In *Origin of Species* Darwin generally referred to 'descent with modification'. Equally, he did not at first use what ultimately became the most famous phrase of all, 'survival of the fittest'. This was coined a few years afterwards, by Herbert Spencer in 1864, after which Wallace suggested Darwin should substitute it for 'natural selection'. All these verbal ambiguities would lead readers in directions that Darwin did not intend. It is not clear from his remaining manuscripts how far he was even aware of the full extent of the difficulties.

Hard on the heels of natural selection, came one further

notion, the new idea he called the 'principle of divergence'. This principle was quick enough to characterize. He said that it was always advantageous for living beings to diversify: 'The more diversified the descendants from any one species become in structure, constitution and habits, by so much will they be better enabled to seize on many and widely diversified places in the polity of nature, and so be enabled to increase in numbers.'[8] Competition for the same 'places' in nature (niches) forced animals and plants to specialize, which in turn stimulated a multiplication of places and greater efficiency in the use of resources. In a worryingly brutal phrase he went on to liken individual animals and plants to steel wedges thrusting ever harder into the softly yielding face of nature. Here lay the roots of some of the harshest economic and social doctrines that would take shape from his writings. Darwin shattered all previous images of pastoral harmony. In his world, the urge to succeed was brutal. Individuals needed to kill to survive.

In explaining divergence for his book, Darwin also introduced one of the most powerful and lasting metaphors of his career. He characterized the history of living beings as a tree, describing extinct ancestral forms as if they were the roots and trunk, each main group of organisms as the branches, and all the multitude of species in existence at the present day as the green leaves and buds: a smoothly spreading evolutionary tree that linked nature and history into a single indivisible living whole, spanning the ages. 'The great tree of life', he declared, 'which fills with its dead and broken branches the crust of the earth, and covers the surface with its ever branching and beautiful ramifications.' His ability to visualize the evolution of life in this way became almost synonymous with understanding it. He made his point with a diagram – the only diagram in the book – which he called 'an odd looking affair but indispensable to show the nature of the very complex affinities of past and present animals'. This showed how a number of ancestral forms might diverge over time, some becoming extinct and others contributing to the next generation – the stark dotted lines hardly indicating the luscious pictures of trees that would soon cascade from naturalists' pens. At the deepest, most satisfyingly symbolic level, Darwin replaced the ancient imagery of the tree of knowledge, the tree of life, with something similar. His tree was time. It was history. It was knowledge. It was life. But it was not divine.

With the core of his theory set out, Darwin let the book sweep onward through a wide range of biological topics. Embryology became intelligible: 'Embryology rises greatly in interest, when we thus look at the embryo as a picture, more or less obscured, of the common parent-form of each great class of animals'. Darwin was proud of this part of his argument and careful to make sure he got it right. He asked his new friend Thomas Henry Huxley (1825–95) to read the chapter before publication. 'The facts seem to me to come out very strong for mutability of species,' he told Hooker when discussing the same chapter.[9] Palaeontology, comparative anatomy and taxonomy would also be transformed, he wrote in anticipation. The anatomical affinities and groupings sought by taxonomists were not just abstract notions, he said, nor were they the physical expression of some divine plan drawn up by the creator, as renowned naturalists like Louis Agassiz or Richard Owen suggested. Instead, the resemblances were caused by genuine blood relationships. Vestigial organs like the appendix in human beings were explained as anatomical remnants left over by history. To Darwin it seemed unlikely that a divine architect would deliberately create such wasteful, purposeless features.

Similarly, the geographical patterns and relationships that plants and animals traced over the globe could be explained on the grounds that species for the most part spread and change. The practical naturalist in him emerged and spoke plainly – the barnacle scholar, the pigeon-lover, the plant experimenter and *Beagle* collector, the traveller at last approaching his goal. Much of the theory's value, he argued, lay in the way it explained and united so many different aspects of the natural world.

Most important in many eyes was his chapter on difficulties. Including such a chapter was an adroit step. In this Darwin discussed many of the problems that would immediately enter a reader's mind, such as the absence of intermediary stages in the fossil record or the unknown mechanisms that might allow the inheritance of mental traits like instincts and the difficulty of envisaging the gradual emergence of complex organs like the eye. Darwin himself had worried endlessly over the same problems. 'The eye to this day gives me a cold shudder,' he confessed to his close friend, the American botanist Asa Gray in 1860.[10] The lack of intermediate forms in the fossil record, for example, was a real puzzler, only explained by what philosophers call a negative argument. He claimed that such organisms would be so rare and transitory, and geological preservation so infrequent and accidental, that it would be highly unlikely to find specimens. Their absence, he stated, could not legitimately overturn his theory. As it happens, he was correct in this surmise. Even with the discovery of fossils like the *Archaeopteryx*, a bird-like reptile in the Solnhofen limestones of Germany, now recognized as a genuine intermediary, the incidence of missing links is still very limited.

This chapter on difficulties was welcomed by reviewers for its honesty. Nevertheless, it was also strategically significant. Darwin chose to write only about the 'difficulties' that he could answer, however tentatively. The difficulties were all of a biological nature. He expected a barrage of factual challenges and provided his answers straight away.

Deliberately, he omitted the two issues that would have occurred to everybody. He avoided any discussion of what evolutionary theory might have to say about human origins, and he sidestepped any debate about a divine presence in the natural world. He remembered the bitter disputes over *Vestiges*. No matter how seriously and cautiously he might treat evolutionary questions he knew that anything he said was bound to ignite furious controversy. So in this book, he was completely silent on the subject of human origins, although he did refer in several places to mankind as an example of specific biological details. Not wishing to appear too revolutionary, however, or openly to attack the cherished beliefs of the faithful, he remarked in the conclusion that, if his views were accepted, 'light will be thrown on the origin of man and his history'.[11]

Similarly he purposefully avoided the first origin of life. He had no systematic history of beginnings to offer, no primeval soup or creative spark. At the end of his book he did mention the likelihood of all ancestral organisms originating in one primordial form. Such ancient origins, he privately believed, were lost in the mists of time and were essentially irreclaimable. When he needed to, he spoke cautiously of the creator, aware that his book might otherwise be labelled subversive. But he was careful not to allow the creator any active role in subsequent biological proceedings. In the first edition of *Origin of Species* Darwin mentioned the origin of this one primordial form as if it were an entirely natural process. In the second edition he used more obviously religious terminology, including an anonymous comment, in actuality made to him in a letter by the Reverend Charles Kingsley, that it was possible to conceive of a creator who allowed species to 'make themselves'; and that the first organic forms had acquired life from the 'breath of the creator'.[12] He evidently did not wish to be perceived as an atheist. For a book that would claim in its title to address the origin of species, Darwin's text in fact refused to propose any theory of absolute origins.

By the end, he had set out one of the most densely impressive proposals of the century. Although in the first edition he did not compare his work directly with those who had gone before, his theory was none the less distinctive. He differed from Lamarck, and his evolutionary grandfather Dr Erasmus Darwin, in that he steered clear of any doctrine of necessary progression or inner striving towards perfection. While Darwin cautiously made space in his scheme for some direct effect of the environment on organisms – the inheritance of acquired characteristics that was popularly assumed to be the main feature of Lamarck's system – the chief difference between them was that Darwin did not allow his organisms any future goal, no teleology or divine power pulling them forwards, no internal effort or act of will that might drive the adaptive changes in specific directions. In Darwin's view organisms shifted randomly. A well-adapted organism might be extremely simple. An insect was as wonderfully adapted as a man.

More significantly he felt sure that he differed from Robert Chambers, the anonymous Mr *Vestiges*, in the solidity of his information, the tightly organized and well-developed theory of change, and his decision to limit the book's scope to one restricted problem and not deal with grand questions of the evolution of the universe, the first sparks of life or the future of the human mind. This certainly made his book dull in comparison with *Vestiges*. But in return it gave him superior standing in scientific circles. Tellingly, he differed most notably from Chambers in putting his name to the text. *On the Origin of Species* was issued with the name of an author on the title page, an author already established as an accredited expert in the field and whose intellectual standing was made plain by the initials of his Cambridge degree and membership of learned societies. The same factors also went some way towards establishing his social and educational difference from Wallace.

There could be no mistaking the weight of thought that lay behind every word, the judicious strategies, the powerful, transformative metaphors, the interlocking double-punch of detail and breadth of vision. Although he subsequently complained that he had been rushed into *Origin of Species*, that it was nothing but an abstract, that his evidence was truncated, and his footnotes and sources were omitted, the book was undeniably Darwin's masterpiece.

'When the views entertained in this volume on the origin of species, or when analogous views are generally admitted, we can dimly foresee that there will be a considerable revolution in natural history,' he declared fervently in the closing pages. 'I look with confidence to the future, to young and rising naturalists, who will be able to view both sides of the question with impartiality.'

> When we no longer look at an organic being as a savage looks at a ship, as at something wholly beyond his comprehension; when we regard every production of nature as one which has had a history; when we contemplate every complex structure and instinct as the summing up of many

> contrivances, each useful to the possessor, nearly in the same way as when we look at any great mechanical invention as the summing up of the labour, the experience, the reason, and even the blunders of numerous workmen; when we thus view each organic being, how far more interesting, I speak from experience, will the study of natural history become![13]

All his hopes came to a crescendo. One particularly attractive spot that he visited during walks with Emma in the countryside around Down House filled his mind.

> It is interesting to contemplate an entangled bank, clothed with many plants of many kinds, with birds singing on the bushes, with various insects flitting about, and with worms crawling through the damp earth, and to reflect that these elaborately constructed forms, so different from each other, and dependent on each other in so complex a manner, have all been produced by laws acting around us... There is a grandeur in this view of life, with its several powers, having been originally breathed into a few forms or into one; and that whilst this planet has gone cycling on according to the fixed law of gravity, from so simple a beginning endless forms most beautiful and most wonderful have been, and are being, evolved.[14]

He hardly anticipated how austere, tragic, dangerous and supremely beautiful his work would appear to others.

Yet who would publish such a book? Hesitantly, in the early months of 1859, Darwin asked Lyell if John Murray might be interested, the same John Murray who published Lyell's books and who in 1845 had issued the second edition of Darwin's *Journal of Researches*. Murray was ideal for several reasons. He and Darwin had enjoyed a businesslike relationship over the *Journal of Researches*. Murray was interested in science, especially geology and chemistry, and well accustomed to initiating shrewd publishing moves like the Home and Colonial Library, a series of edifying works for the middle classes, and the famous *Handbooks*, the first holiday guidebooks for Victorians, predating Baedekers by a few years.

More than this, Murray was rapidly becoming one of the most important scientific publishers of the Victorian era. His doors in Albemarle Street, in the centre of literary London, were open to authors of all shades of opinion. Murray offered a contract and Darwin gratefully accepted, the start of a relationship that lasted for the rest of his life.

The constant writing, however, was eating away at Darwin's health. 'My God how I long for my stomach's sake to wash my hands of it – for at least one long spell,' he complained. 'I am becoming as weak as a child,' he groaned to Hooker, 'miserably unwell & shattered.' The summer of 1859 passed in a turmoil of proof-reading. All Darwin's doubts about his writing style returned with a vengeance. 'There seems to be a sort of fatality in my mind leading me to put at first my statement and proposition in a wrong or awkward form', he reflected afterwards.[15] Emma Darwin helped whenever she could. She read the *Origin* in full during the proof stage and loyally tried to help her husband convey his thoughts accurately to readers. There is no evidence that she tried to censor his text. On the contrary, the two of them discussed awkward sentences in the evenings until they found a form that captured what he was really trying to say, and she would tease him about his poor use of commas. Lyell read the proofs while he travelled around the continent on his summer holidays.

At the last minute Darwin adjusted the title according to Murray's recommendation. Darwin's first suggestion was apparently too complicated: 'An Abstract of an Essay on the Origin of Species and Varieties through Natural Selection'. Common sense suggested to Murray that the words 'abstract', 'essay' and 'varieties' should go, and that 'natural selection', a term with which Murray thought the public would not be familiar, ought to be explained. The agreed title was hardly less cumbersome: *On the Origin of Species by means of Natural Selection, or the Preservation of Favoured Races in the Struggle for Life.*

He was in an absorbed, slavish, overworked state, he told William Darwin Fox in a letter. 'My abominable volume... has cost me so much labour that I almost hate it.' Sometimes he recoiled from seeing nature the way his selection theory demanded. 'What a book a Devil's chaplain might write on the clumsy, wasteful, blundering, low & horridly cruel works of nature!' he once exclaimed to Hooker.[16] 'I have been so wearied and exhausted of late,' he complained in September 1859. 'I have for months doubted whether I have not been throwing away time & labour for nothing.'

Then, on 1 October 1859, he recorded in his diary, 'Finished proofs', and calculated that the whole process had taken thirteen months and ten days from start to finish. On 2 October he left Down House, exhausted and sickly, and made his way to a water cure establishment in Ilkley, at the foot of the Yorkshire moors. 'I am worn out and must have rest... Hydropathy and rest – perhaps that will make a man of me.'[17]

His book was published in London on 24 November 1859. Darwin was in Ilkley on the day of publication, returning home a fortnight later.

CHAPTER 4

Controversy

The tidal wave of comment began almost immediately. Despite all Darwin's carefully amassed evidence, and his repeated invitations to the reader to consider the issue impartially, Victorians found it nearly impossible to accept the idea of gradual change in animals and plants, and equally hard to displace God from the creative process. Yet this volume, and the ensuing debate, placed the issue of evolution before the public in a form that could not be ignored. The essence of Darwin's proposal was that living beings should not be regarded as the carefully constructed creations of a divine authority but as the products of entirely natural processes. As might be expected, there were scientific, theological and philosophical objections from all quarters, often mixed up together. Were human beings to be included? Should science be allowed to address questions that up until then were the business of theologians? What was the purpose of our world if there were no reason for the existence of virtue? How could an ape be my grandfather?

Journalists, men of letters, merchants, businessmen, educators and ordinary men and women added their voices to the throng. Bishops, poets, kennel-hands and governesses read the book. Even Queen Victoria took an interest, although she confided to her daughter that she expected it would be too difficult to understand. Nor was the reaction confined to Britain. In France, Germany, Italy, Sweden, Russia and North America, and progressively all over the globe, people from every walk of life discussed the idea of evolution by natural selection and relocated this controversial issue within their own cultural contexts. It was one of the first genuinely public debates about science to stretch across general society. These varied responses, evocative of the cultural diversity of the nineteenth century, remind us that the introduction of new ideas is rarely straightforward and that the past histories of science have involved many different forms of publication, many different audiences and many different languages as well as the ideas themselves.

In retrospect, it is also evident that *Origin of Species* contributed markedly to other fundamental shifts already under way in the West – in particular in religious affairs. The Anglican Church still lay at the heart of the British nation's daily life and provided the framework in which most people operated, either more or less devoutly according to private inclination. Yet its grip was looser than before. Schisms and fractures appeared, splinter groups broke away, dissatisfaction was expressed. Dissenting and non-conformist groups claimed the right to worship in their own manner, to educate the young, to be represented in parliament, to take public positions and have their views heard. A non-denominational

University College was established in London, rapidly filling up with the brightest and best unconventional minds. A number of theologians converted to Catholicism. Prominent men and women among the establishment declared themselves sceptics or critics of traditional doctrine. Even clergymen undermined their own message. One of the authors of *Essays and Reviews* was the Reverend Baden Powell (grandfather of the scoutmaster), professor of geometry at Oxford University, who once claimed miracles could not occur, praised Darwin's *Origin* as 'a masterly volume' and pronounced in favour of 'the grand principle of the self-evolving powers of nature'. A book such as *Origin of Species* was therefore bound to arouse hot controversy and range more widely than science itself. Discussion was never going to be limited to butterflies or primroses.

Surprising as it may seem, there was little sustained opposition to Darwin's book on the grounds that it directly challenged the account of creation in Genesis. Learned biblical study since the Enlightenment had encouraged Christians increasingly to regard the early stories as potent metaphors rather than literal accounts. Biblical fundamentalism is mostly a modern concern, not a Victorian one. The real challenge of Darwinism for Victorians was that it turned life into an amoral chaos displaying no evidence of a divine authority or any sense of purpose or design.

This was a political and social issue as well as a theological one. The reaction of many respectable middle-class believers was to reject evolution because it threatened the

Church's role in guarding the nation's morals and social stability. Some freethinkers moved in the opposite direction and used evolution to level varying degrees of criticism at ecclesiastical policy and the state. A few hardliners already well on the way to atheism abandoned religious faith altogether. Equally hard Calvinists managed to accommodate the idea of natural selection by integrating it with humanity's struggle to overcome sinfulness. But there were also liberal Christians willing to accept evolution as a fact of nature if it could be reconciled with moral principles. Perhaps evolution could be regarded as a purposeful process regulated by God? This compromise soon emerged in Britain. In 1861 the astronomer John Herschel wrote that he could believe in an 'Intelligence' that guided the steps of change according to the laws of science. By the end of the century a number of Anglican clergymen, such as Charles Kingsley and Frederick Temple, promoted a similar theology in which the shaping of the earth and its living beings were seen as a continuous process controlled by laws God had instituted in the beginning. Or again, was it possible to replace the mechanical process of natural selection with something else of divine origin? A number of scientists, among them Darwin's friend Asa Gray, took this route and put back the moral purpose and future goals – the teleology – that Darwin had removed.

One of the most well-known aspects of the *Origin of Species* controversy is that Darwin kept out of the limelight. On the face of it, this is completely true. Darwin never enjoyed public debate, hated confrontations in which his honour or honesty might be called into question, preferred to stay quietly at home in the background, and was content to let others wave the flag more vigorously than he felt able to do himself. Privately, he believed that disagreements between scientists were generally fruitless. None the less the underlying story is more complex. Darwin kept in close touch. Even though he stayed put at Down House, a barrage of correspondence was despatched and received daily. His letters were out there in the world of argument: encouraging, supporting, nudging, explaining, politely disagreeing, thanking, consulting and advising. He used letters to persuade and to influence. He used them to get favourable reviews, correct mistakes, arrange translations and produce revised editions. He gathered support, made new contacts, found out things. Without this extraordinary correspondence, rising to a peak of some 500 letters a year after *Origin of Species* was published, Darwin's theory would have sunk. In this he was materially helped by the rapid development of the Victorian postal system, brought to a peak of efficiency by Rowland Hill from the 1840s and 1850s, and the expanding infrastructure of empire.

Scholars agree that the course of the *Origin* controversy was unique in several respects. The book's wide and immediate impact in Britain was greatly enhanced by an expanding publishing industry and new review journals being produced for rapidly diversifying audiences. It was greatly enhanced, too, by mid-century peace and prosperity, political stability and imperial expansion. The audience for science was the largest and most appreciative that it had ever been, its appetite whetted by the development of local scientific societies, lending libraries, public lectures and exciting practical demonstrations of electricity, chemistry or magnetism, and reinforced by the broadening availability of manufactured goods and obvious achievements in roads, railways, bridges, ships and canals. Writings like Chambers's *Vestiges* and Tennyson's *In Memoriam* already helped readers explore the big issues of human existence, questions of origin, meaning and purpose.

Highly characteristic, too, was the personal element. Four of Darwin's friends carried the brunt of the public storm: each one an acknowledged specialist in his scientific field, independent, clever and far from sycophantic. These four supported Darwin wholeheartedly even while pointing out flaws in his evidence or reasoning. They stood united, gathering their own disciples and followers, engaging in individualized battles on Darwin's behalf but also moving the debate further and wider, drawing in other thinkers, other topics, other implications, in an incremental process that ultimately generated major transformations in cultural attitudes and scientific thought. With Darwin busy in the background writing letters, these four recruited a standing army, commandeered the journals, invaded the learned societies, monitored the universities, dominated dinner parties and penetrated the byways of empire. The opposition never quite consolidated in the same way. There were of course individual heavyweight opponents who publicly challenged

Darwinism, some of them witheringly effective, but no united group rallied to the attack or mustered behind powerful spokesmen. There was never an explicit anti-Darwinian movement, in the same way as there was a pro-Darwinian group held together by intellectual commitment and friendship. The existence of this Darwinian alliance was perhaps the single most important feature of the debate and contributed markedly to the ultimate triumph of evolutionary theory. At its core were Charles Lyell, Joseph Hooker, Asa Gray and Thomas Henry Huxley.

Inspired by Darwin's ideas, Lyell focused on human archaeology and prehistory in an impressive text called *The Antiquity of Man* (1863). In this book he undermined the traditional story of the Creation and Flood, and showed how humans had appeared on the globe much earlier than anyone then thought possible, contemporaneously with animals now only known as fossils. Though he did not coin the expression 'cave man', which came later, nor could he claim to be the only one intrigued by discoveries of worked flints and stone arrowheads, Lyell was among the first to write about these early peoples within a broadly evolutionary structure. Since he was one of the most widely respected scientific authors of the century, the effect of extending Darwin's thesis into the prehistoric world was incalculable. Privately, he felt unable to go as far as Darwin in believing that human beings were entirely natural organisms. To the end of his life he felt that humans possessed a divine soul. Once he told Huxley that he 'could not go the whole orang'.[1] Lyell's interest in early human cultures was soon extended by a generation of gifted evolutionary anthropological thinkers. John Lubbock, a younger friend and neighbour of Darwin's, discussed the archaeological evidence of primitive cultures in Europe in his studies *Pre-historic Times* (1865) and *The Origin of Civilisation* (1870). This was powerfully followed by the main thrust of high Victorian cultural anthropology in Edward B. Tylor's evolutionary *Researches into the Early History of Mankind and the Development of Civilization* (1870), Lewis Henry Morgan's *Ancient Society* (1877) and Sir John Evans's work. These men codified the late nineteenth-century belief that human development had progressed through a sequence of stages from savagery through barbarism to civilization, and that primitives were relics of the earliest stages that could be studied for insights into the history of mankind.

While Lyell grappled with early humanity, Joseph Hooker aimed at the empire of botany. Hooker's father – and then Hooker himself – was director of Kew Gardens, located just outside London, the largest and fastest-growing centre for botanical research in the world, with a special focus on economic botany and colonial expansion. Hooker's public work at Kew furthered the introduction of plantation crops in far-flung corners of the globe such as tea, coffee, sisal, sugar, mahogany, cinchona, cotton and flax. Much neglected by historians, botany during the nineteenth century was the most significant science of its day, creating and destroying colonial cash crops according to government policy and building the economic prosperity of a nation. Almost single-handedly Hooker coordinated the activities of British colonial gardens and orchestrated a worldwide correspondence with other botanists. Like Lyell, he was one of Darwin's closest friends, a man that Darwin trusted and liked. He was the first to show how Darwin's theories might work in the plant world and supported him loyally in publications, reviews and correspondence. He never wrote a signature book like Lyell, but his influence and scientific position at the centre of imperial science was a key strength on Darwin's side.

On the other side of the Atlantic, Asa Gray defended Darwin just as effectively. Based at Harvard University in Cambridge, Massachusetts, Gray was also a botanist, a rival professor to Louis Agassiz, the most celebrated naturalist in the United States. Agassiz was no fan of *Origin of Species*. His belief that the 'essence' of every species should reflect God's divine blueprint guaranteed that he would emphatically reject evolution – how could one hope to classify anything if it was constantly changing? Gray and Agassiz argued fiercely about Darwinism in public meetings in Boston in 1859 and 1860, and perhaps Gray was the only man in America who could (occasionally) get the better of Agassiz in debate. Gray felt, however, that Darwin's scheme should be modified to help those who sincerely believed in God's presence in the natural world. To him, natural selection, acting blindly on occasional chance variations, did not seem sufficient to account for so many organisms exquisitely 'designed' for their role in life. Gray therefore proposed that God created good and useful variations which natural selection then preserved in a population. While this view was completely antithetical to Darwin's proposal, Gray promoted it earnestly in several widely distributed reviews. Darwin respected Gray's opinion, saying that it was the best natural theological commentary he had ever read. He appreciatively declared that every one of Gray's attacks on opponents 'tells like a 32-pound shot'. Soon he was convinced that 'no other person understands me so thoroughly as Asa Gray. If I ever doubt what I mean myself, I think I shall ask him!'[2]

Last, and most famous of all, Thomas Henry Huxley, the brilliant zoologist and comparative anatomist, cast himself as 'Darwin's bulldog'. Flamboyantly, he defended Darwin on the question of ape ancestry and the close anatomical resemblance between humans and primates, and reigned supreme over what can justly be called the marketing of evolutionary theory – a heady publicity campaign for a new kind of science based on rational thought untainted by religious belief. One important plank of his platform was to wrest education from the hands of the clergy, for schoolchildren and university students were for the most part still educated within traditional Anglican institutions or by dissenting church missions. Another was an increasingly violent feud with the rival comparative anatomist Richard Owen, a man profoundly against evolution, whom Huxley thought was blocking his path to success. Superintendent of the natural history collections at the British Museum (at that time located in the Bloomsbury building), and in many eyes the leading naturalist in Britain, Owen made a brutal attack on *Origin of Species* in the *Edinburgh Review* of April 1860 that angered Darwin and supplied the backdrop for much of Huxley's venom during the early 1860s. Other personal characteristics included Huxley's intense dislike for religious 'claptrap' and zest for public showdowns. Late in life he was credited with having coined the word 'agnostic' to describe his position: one who cannot believe without rational evidence for that belief. 'In matters of the intellect, do not pretend that conclusions are certain which are not demonstrated or demonstrable... That which is unproved today may be proved, by the help of new discoveries, tomorrow.'[3] His reputation for wit and coruscating prose was well established.

In fact, the months immediately surrounding publication of the *Origin of Species* really belonged to Huxley. As Huxley recalled it, the beauty of Darwin's theory flashed on him like lightning showing the way home. 'How extremely stupid not to have thought of that!' he exclaimed. He composed three magnificent reviews, one in *The Times* newspaper, the others in well-known literary journals, the *Westminster Review* and *Macmillan's Magazine*.

In the *Westminster Review* he issued a battle cry. *Origin of Species* was a 'veritable Whitworth gun in the armoury of liberalism'. It would free the world from theological dogma:

> What is the history of every science but the elimination of the notion of mystery or creative interferences?... Extinguished theologians lie about the cradle of every science as the stran-

> gled snakes besides that of Hercules, and history records that whenever science and dogmatism have been fairly opposed, the latter has been forced to retire from the lists, bleeding and crushed, if not annihilated; scotched if not slain.[4]

This opening blow against religion not only made Huxley's name but served him and Darwinism well in the future.

Huxley's first public showdown – now regarded as a famous set-piece in the history of science – took place at a meeting of the British Association for the Advancement of Science at Oxford in June 1860. There are few records left of the occasion. No one even knew for sure who had won. Nevertheless, the occasion meant a great deal in historical terms. It became an enduring symbol of an angry clash between science and religion over the origin of species.

As was customary, the British Association meeting ran for a week in the summer and made the latest developments in science more widely known to the public. Irresistibly drawn by the prospect of heated exchanges about monkey-ancestors, an unusually large number of people arrived at the session held in the Oxford University Natural History Museum on Saturday 30 June. Darwin did not attend the meeting because he was ill. Huxley and Owen were both there. Earlier in the week there had been several intellectual skirmishes between them, especially when Owen asserted that there was no anatomical evidence in primate brains for evolution. Huxley had jeered at his competence. 'You and your book forthwith became the topics of the day,' Joseph Hooker told Darwin.

The session promised sparks. An American philosopher John William Draper, known for his denunciations of Roman Catholicism, was scheduled to speak on the evolution of human society 'with reference to the views of Mr. Darwin'. As it turned out, Draper's talk was dreary. The mood visibly lightened when Bishop Samuel Wilberforce, the current Bishop of Oxford, rose to speak. Wilberforce was a powerful orator, witty and eloquent. As a theologian, he naturally used the occasion to defend the divine creation of humankind. He had just written a damning review of Darwin's book for the *Quarterly Review* and his speech repeated many of the points published there, particularly using anatomical information supplied by Owen. How could anyone seriously believe that mankind had developed from oysters, he asked? At some point he turned to Huxley and facetiously enquired 'was Huxley related to an ape on his grandfather's or grandmother's side?'

The audience smelled blood. So did Huxley. He answered at length, first repudiating the anatomical arguments used by Wilberforce and then praising the way that Darwin's theory united previously chaotic data. The exact words he used were not recorded. But his final thrust was to say that he 'would rather have a miserable ape for a grandfather than... a man who introduced ridicule into a grave scientific discussion'. The audience cheered and went away convinced that Huxley would rather have an ape for a grandfather than a bishop. They felt they had witnessed in miniature a titanic confrontation between the church and science – two utterly incompatible views on the position of mankind in the natural world.

Afterwards, Huxley made his position clear in a small, vivid volume called *Evidence as to Man's Place in Nature* (1863), a popular book that addressed audiences primarily wanting to hear about apish ancestors. It included a lucid and favourable exposition of Darwin's theory. Here Huxley continued his argument with Richard Owen by attacking Owen's anatomical work on the great apes. For a long time, Owen had insisted there was a small fold in the membranes at the base of the human brain (the hippocampus minor) that could not be found in any of the apes. This, Owen thought, along with other differences such as the human hand and upright posture, indicated the special nature of human beings. Huxley vehemently disagreed. Professional reputations and expertise were at stake here. Simple observation would not be able to resolve the issue because the disagreement rested on questions of judgement, interpretation and scale. In his book Huxley claimed that there were clear continuities in anatomy between gibbons, gorillas and mankind. Visual reinforcement was supplied in an engraving that showed the skeletons of four species of ape lined up in an evolutionary sequence with a human being. This first pictorial representation of evolution has since become as iconic as the double helix of DNA.

Huxley's view has come to prevail. At the time, however, his argument with Owen raged through the popular press bringing the shocking possibility of ape ancestry home to the masses. Charles Kingsley found the clash a rich source of satire when writing his children's book *The Water Babies* in 1863. He included caricatures of Huxley and Owen quarrelling over the definition of a water baby, and joked that 'apes have hippopotamus majors in their brains just as men have... Nothing is to be depended on but the great hippopotamus test'. Edward Linley Sambourne, the artist, illustrated the two men quarrelling over a baby in bottle.

All the while apes were pushing noisily to the fore. Most remarkably, the gorilla suddenly became front-page news through the exploits of Paul Du Chaillu, an explorer who in 1861 brought specimens and skins to Europe. At least one of those skins (perhaps as many as three) was stuffed and travelled about with Du Chaillu as he gave public lectures about the gorilla's ferocity and the dangers he escaped during his West African travels. Few had ever seen or heard anything like it before – gorillas were almost entirely unknown in the West until 1854 when bones were despatched from Africa to Harvard University for identification. Victorians were horrified to think that these reputedly violent animals – distorted men in shape and size, representing the brutish, dark side of humanity – were possible ancestors. Museum curators competed shamelessly for the carcasses, until Owen persuaded the trustees of the British Museum to pay a fortune to acquire six skins from Du Chaillu. Elsewhere, humorous journals such as *Punch* seized on the idea of apish grandfathers and printed a wide variety of cartoons and satires depicting humanized gorillas. 'Am I a Man and a Brother?' asked an ape in one famous illustration in *Punch*, May 1861, playing simultaneously on Du Chaillu's stuffed gorilla and the anti-slavery movement. In truth, the furore generated by evolutionary ideas pulled apes, anatomy, polemic, fear, disgust and sensationalism into a single debate. Benjamin Disraeli, the future Conservative prime minister, exposed the unease of his contemporaries in 1864 when he asked 'Is man an ape or an angel?' He went on to assure his audience that he was on the side of the angels.

Others engaged with the *Origin*'s arguments with philosophical curiosity. While John Herschel complained that natural selection was the law of 'higgledy-piggledy', and that Darwin was not following traditional procedures of demonstration and proof, Henry Fawcett at Cambridge University and the philosopher John Stuart Mill compared the new style of reasoning favourably against the old. Mill endorsed Darwin's work in the 1862 edition of his *System of Logic*, saying that although Darwin had not proved the truth of his doctrine, he had shown that it might be true, an 'unimpeachable example of a legitimate hypothesis... He has opened a path of inquiry full of promise, the results of which none can foresee'.[5] Ernest Renan, the noted theological writer, whose *Life of Jesus* deliberately left out the divine, said much the same thing. So did George Henry Lewes when discussing natural selection in his *Animal Life* (1862): 'it *may* be true but we cannot say that it *is* true'. These thoughtful authors saw the argument's explanatory value and were not prepared to dismiss it out of hand simply for religious reasons.

Even those who disagreed with Darwin were mostly able to concede the merits of his case. The great philologist Friedrich Max Müller addressed Darwin's theories in lectures about the origin of speech during the winter lecture season in London, 1861–2. Müller forced his audience of fashionable swells to think carefully about what it was to be human. Had our gift of language developed from animal sounds? He thought not. Words could only exist with thoughts, and thoughts were the special preserve of humans. Animals did not have anything like human concepts, he claimed. Müller vigorously opposed evolutionary theory. Yet he praised the notion of natural selection and applied it enthusiastically to the descent and historical relationships of Indo-European languages, as the other great language scholar of the day, August Schleicher, was to appreciate.

Poets and authors were not far behind. Alfred Lord Tennyson never accepted Darwin's proposals but was keen to meet him in 1868 when they both coincided on holiday in the Isle of Wight. Tennyson had been deeply affected by Chambers's *Vestiges* and did not trouble to distinguish the two books from each other. 'Darwinism, man from ape, would that really make any difference? Time is nothing, are we not all part of deity?' he remarked to William Allingham in 1863. Tennyson's gloom about the void after death, although not generated by Darwinism, nevertheless moved him broadly in the same direction as *Origin of Species*. Robert Browning similarly questioned whether there was any purpose in human existence. But perhaps Matthew

Arnold spoke clearest of all for Victorians beset by religious doubt. In his poem 'Dover Beach' (1851) the sea of faith that once supported spirituality was now nothing more than a 'melancholy, long, withdrawing roar'.

And Karl Marx was famously intrigued by Darwin's thesis, saying on several different occasions that he saw in its workings the capitalist system of competition and laissez-faire. At one time it was thought that Marx wanted to dedicate *Das Kapital* to Darwin but this was based on a misunderstanding. Marx certainly mentioned *Origin* in his text and sent a presentation copy of his third edition of *Das Kapital* to Darwin as a mark of respect. It remains in Darwin's book collection with an inscription from Marx inside. The confusion emerged from a misidentification of a letter to Darwin. The letter was actually from Edward Aveling, the political philosopher and Marx's son-in-law, who enthusiastically adopted Darwin's secular insights. Aveling asked if Darwin would accept a dedication in one of Aveling's books. Not wishing to be publicly associated with Aveling's atheism, Darwin rejected the request.

Running alongside these intense debates over apes and angels were two main scientific objections to *Origin of Species*. The first cut to the heart of Darwin's proposal and queried the origin and preservation of favourable variations. The theme was picked up in 1867 by Fleeming Jenkin, a Scottish engineer and friend of Robert Louis Stevenson. Jenkin asked how could advantageous individuals survive and reproduce in sufficient numbers to shift the whole population in the same favourable direction? Jenkin was hampered, like many of his contemporaries, by believing in what was then known as 'blending inheritance', where the characteristics of any two parents were thought to mix and blend in the offspring. If this was so, then any favourable new traits would be blended out in future generations. It was only later, with Moritz Wagner's insistence on geographical isolation in the evolutionary process (a notion itself developed from Darwin's work), that the blending problem looked as if it was solved.

Darwin was very perplexed to answer the point satisfactorily. He recognized – indeed the majority of reviewers told him – that the major gap in his book was that he did not explain the origin of variations nor the process of heredity. He tried to do so in his next significant book, *On the Variation of Animals and Plants Under Domestication* (1868). He devised a theory of inheritance that he called 'pangenesis' in which each part of the parent's body was thought to throw off minute particles, or 'gemmules', which accumulated in the sexual organs to be transmitted in reproduction. Parental gemmules did not blend, he claimed. Instead they were reorganized.

The scheme was roundly criticized, first by Huxley, and then most tellingly by Darwin's cousin Francis Galton (1822–1911), an enthusiast for evolution who was interested in inheritance and the 'fitness' of the human race, and took Darwin's theories into the human domain under the label of eugenics. Galton hoped to prove Darwin's pangenesis by making blood transfusions between rabbits and then breeding from them, but, to his dismay, ended up showing that gemmules were not present in the blood. The two cousins never saw eye to eye over the business. Late in life Darwin was gratified when the pioneer geneticist August Weismann (1834–1914) took up the notion of gemmules (pangenes) as a vehicle for the transmission of information from parent to offspring.

Commonly disregarded by historians of genetics, Darwin's theories should perhaps be placed more in the mainstream of investigations into inheritance. He was one of many others who at that time felt that heredity must hold the key to the question of origins. The problem was under prolonged and intense investigation from the 1860s onwards by Charles Naudin, Karl Wilhelm Nägeli, Karl Friedrich Gärtner and Weismann. Coincidentally, Gregor Mendel (1822–84) was also at work in the monastery in Brno (Moravia, now Czech Republic) where he spent his life as a pastor. Mendel's crossing procedures with pure lines of peas and other garden species, although later the foundation of the modern science of genetics, were more or less ignored when published in the local natural history journal in 1865, and there is no evidence that Darwin read Mendel's article or that it would have provided him with the necessary clue if he had done so. Darwin's theory of inheritance did not convince contemporaries, who continued to point to the gap in his argument.

The other scientific sticking point emerged in 1866 when

the experimental physicist William Thomson (later Lord Kelvin) announced that the earth was not old enough for evolution to have taken place. Propelled by anti-evolutionary Scottish Presbyterian inclinations, Thomson stated that 100 million years was all that physics could allow for the whole of the earth's geological history. Uniformitarians like Lyell and Darwin, who believed in slow and gradual changes over vast aeons of time, were dumbfounded. Thomson's arguments were 'an odious spectre'. Decades of continuing debate over the age of the earth were resolved only with the discovery of radioactivity early in the twentieth century that provided a different way of counting and let the evolutionists off the hook.

Darwin responded to these criticisms, and others, in the pages of succeeding editions of his book. One of the most important changes has been noted by many commentators. In the closing pages of the first edition Darwin had written of life being breathed into a few primordial forms. For the second edition he altered this to read 'the breath of the Creator', a concession that he came to regret. In the second edition he also added a few words (unattributed) from a letter written to him by Charles Kingsley, indicating that it was possible to believe in God as the ultimate author of evolution. These words remained more or less intact until later editions. Historians have also remarked on Darwin's willingness to include increasing levels of Lamarckian evolution as the years went by.

During his lifetime Darwin published six editions of

Origin of Species, 18,000 copies in total. The first edition numbered 1,250 copies. It is worth comparing these figures with two of the most popular scientific books of the century, *Vestiges* (1844), which sold 24,000 in sixteen years, and nearly 40,000 by 1890, or George Combe's *Constitution of Man* (1828), which sold 11,000 in eight years. There were eleven translations of *Origin* produced before Darwin's death in 1882, and numerous shortened versions and commentaries, many of which required close cooperation with the authors and editors. It has appeared in a further eighteen languages since.

One area where Darwin's theory obviously impinged on human society was in the suggestion that there was a struggle for existence among nations and races. After the *Origin of Species* was published, the notorious doctrine of 'social Darwinism' took the idea of success to justify social and economic policies in which struggle was the driving force. Intimately tied up with national economies, embedded in powerful class, racial and gender distinctions, dancing to a variety of political commitments, there was no single form of social Darwinism. Indeed, some scholars argue that it hardly derived from Darwin and Wallace's scheme of natural selection at all but was more closely connected with Herbert Spencer's pervasive social evolutionism. Spencer's nostrum of 'survival of the fittest' was well suited to describe economic expansion, rapid adaptation to circumstance and colonization.

Be this as it may, the dominant economic strategy of

developed nations during the second half of the nineteenth century took shape in the aftermath of the *Origin*'s publication. It was common to use the book directly to legitimize the competition that flourished in free-enterprise Victorian capitalism. Darwin was perfectly aware of these activities and may even have approved of them. Early on he noted that a reviewer in Manchester (one of the largest manufacturing cities in Britain) stated that the *Origin* promoted the notion that 'might was right'. Darwin's ideas were welcomed by many industrial magnates and manufacturers. By the end of the century they were being put into action by the businessmen, philanthropists and robber barons who masterminded the development of North American industry, especially J. D. Rockefeller and the railway owner James J. Hill, who used 'survival of the fittest' as their catchphrase. In their view the strongest and most efficient company would naturally dominate the market and stimulate economic progress on the wider scale. Others, like Andrew Carnegie, the émigré Scotsman who created a vast fortune and spent the rest of his life giving it away, revered Spencer. These commitments were heavily biased towards the political right. Few such thinkers believed in socialism or state support for the poor. A welfare state or subsidized industry, it was assumed, would encourage idleness and permit an increasing number of 'unfit' people or firms to survive, thereby undermining economic and social progress and national health – an obvious resurgence of Malthus's original ideas, now poured back into economic thought with a fully 'scientific' backing provided by Darwin.

Enthusiasm for free enterprise merged readily into growing ideologies of imperialism and eugenics. The 'survival of the fittest' supported notions of inbuilt 'racial' difference and appeared to vindicate harsh and continuing fights for territory and political power on the international stage. The success of white Europeans in conquering and settling in Tasmania, for example, seemed to 'make natural' the wholescale extermination of Tasmanian aboriginals. Conquest was deemed a necessary part of progress. A fairly typical view was expressed by Karl Pearson (1857–1936), the committed Darwinian biologist and London statistician. No one should regret, he said in 1900, that 'a capable and stalwart race of white men should replace a dark-skinned tribe which can neither utilize its land for the full benefit of mankind, nor contribute its quota to the common stock of human knowledge'.[6]

Social commentators appeared to agree. Eugenics was given its name and leading principles by Francis Galton in the 1880s, drawing on nationalistic, racial and social assumptions already well established but acquiring great social force when attached to evolutionary theory. Galton felt that civilized societies tended generally to prevent natural selection working, in the sense that many of the 'unfit' were preserved by medicine, charity, family or religious principles, whereas in a state of nature such people would die. The worst elements of society were the most fecund, he said. The human race would deteriorate, he declared, unless policies were introduced to reduce breeding rates among what he categorized as the poorer, unfit, profligate elements of society and promote higher rates among the worthy middle classes. One of the most pervasive social movements of the early twentieth century, spreading widely through Europe and the Americas, eugenics increasingly became the channel through which anxieties about racial and political decline were projected on to the 'unfit' in society. Many eugenicists believed passionately in improving humanity, in political meritocracies, education, birth control and greater freedom for women, were advocates for technological and scientific advance and often committed socialists, and yet also promoted nationalism, chauvinism and prejudice. While Darwin's *Origin of Species* can hardly account for all the racial stereotyping, nationalist fervour and harshly expressed prejudice to be found in years to come, there can be no denying the impact of providing a biological backing for human warfare and notions of racial superiority.

Towards the end of his life, it could almost be said that the *Origin of Species* devoured Darwin. The constant pressure was draining. Through the 1860s and 1870s he became ill more frequently and for longer periods. One unpleasant episode of sickness dominated 1864, during which Darwin was bedridden for much of the time, vomiting and nauseated, unable to see friends or work except at the most sedentary occupations, too weak even to write his usual cascade of letters. His wife Emma and daughter Henrietta acted as amanuenses. He gave up the water-cure and placed his faith in dietary regimens and resting his 'nerves'. Under the care of several physicians he also took a variety of Victorian remedies for dyspepsia. In 1866, when he re-emerged from the sickroom, he had become the frail old man with the enormous grey beard that everyone remembers.

Yet he managed to write a number of other books following *Origin of Species*. The first was on orchids in 1862, that represented a very deliberate exploration of adaptations in nature, what he called 'a flank move on the enemy'. It was his answer to William Paley's heavenly watchmaker and stimulated much theological discussion with Asa Gray. 'I cannot think that the world, as we see it, is the result of chance; and yet I cannot look at each separate thing as the result of design', he told Gray.

By far the most influential was his *Descent of Man, and Selection in Relation to Sex*, published by John Murray in two volumes in 1871. After all the heated discussion about human origins perhaps it was a little overdue. Other voices, other texts had meanwhile put the case both for and against the animal basis of humankind. Indeed, Darwin confessed to a correspondent that he felt 'taunted with concealing my opinions'. However, he was at last dealing with what he called 'the highest and most interesting problem for the naturalist'. Some material was too extensive even to include in this new book, so Darwin set it aside for an innovative volume published the following year, *The Expression of the Emotions in Man and Animals* (1872). These two titles represent Darwin's great anthropological cycle, his 'man' books, the final, vital counterpart to *Origin of Species*.

Alfred Russel Wallace played an important role in Darwin's thinking in this regard. After Wallace's return to England in 1862 the two had become good friends, each respecting the other's achievement. Nevertheless, Wallace bowed to what must have seemed the inevitable. Much of the shock of evolution had died down by the time of his return. Darwin's *Origin of Species* occupied the front line and the word 'Darwinism' was already circulating as a synonym for evolutionary theory. Wallace's position was therefore different from that of Lyell, Hooker or Huxley, and perhaps uncomfortable, being neither disciple nor primary author. Eventually he wrote one of the best nineteenth-century texts on natural selection, modestly calling it *Darwinism* (1889). Somehow he never gained the celebrity or status in Victorian science that Darwin achieved and is often regarded by historians as an outsider, a fascinating figure who joined the establishment only briefly. Increasingly he and Darwin agreed to differ on particular points. Wallace revealed that he did not like the expression 'natural selection' and in 1868 persuaded Darwin to introduce the expression 'survival of the fittest', taken from Herbert Spencer's writings.

Their main difference was over the origin of human beings. Wallace wrote two compelling articles on human evolution in the 1860s. In the second, published in the 1869 *Quarterly Review*, he declared that natural selection was insufficient to explain all the evolutionary beginnings of humankind. He proposed instead that natural selection pushed our apish ancestors only to the threshold of humanity. At that point, physical evolution stopped and something else took over: the power of mind. The human mind alone continued to advance, human societies emerged, cultural imperatives increased, a mental and moral domain became significant and civilization took shape. Not every society developed at the same rate – primitives were slow, Caucasians fast. For all his genuine social democratic principles, Wallace believed in a hierarchy of savage and civilized. Darwin was taken aback. 'I hope you have not murdered too completely your own and my child,' he exclaimed in surprise. It was partly the impact of seeing this article that encouraged him to express his own views fully in *The Descent of Man*. He was determined to show that everything human – language, morality, religious sense, maternal affection, civilization, appreciation of beauty – had emerged from animals.

The book was large and Darwin solicited assistance from many friends and scholars already working within evolutionary anthropology, such as Huxley or the talented German-speaking naturalists Ernst Haeckel and Carl Vogt. The text included an important new idea that he called 'sexual selection'. This accounted, as he thought, not only for the differences between males and females – the secondary sexual characteristics as they are usually called – but also the differences between human races. Darwin used the terminology of his day, writing of racial characteristics and racial 'types'. He felt certain that sexual selection was 'the main agent in forming the races of man'.

The idea was relatively simple. Animals, he said, possess

many trifling features that are developed only because they contribute to reproductive success. These features have no adaptive or survival value. The classic example is the male peacock that develops large tail feathers to enhance its chances in the mating game even though the same feathers actively impede its ability to fly from predators. The female peahen, argued Darwin, chooses the most well-adorned mate and thereby passes his characteristics on to the next generation. It was a system, he stressed, that depended on individual choice. In *The Descent of Man* Darwin devoted nearly half the book to establishing the existence of this sexual selection in birds, mammals and insects. Wallace disagreed with him on several substantial points, particularly the purpose of protective coloration in birds and insects.

Darwin then extended the idea to explain the divergence of early humans into the racial groups that physical anthropologists described. Preference for certain skin colours was a good example. Early men would choose wives according to local ideas of beauty, he suggested. The skin colour of an entire population would gradually shift as a consequence. 'The strongest and most vigorous men... would generally have been able to select the more attractive women... who would rear on average a greater number of children.'[7] Societies would have dissimilar ideas about what constituted attractiveness and so the physical features of various groups would gradually diverge. In effect, humans would make themselves. The same argument applied to mental characteristics, pushing some groups away from the tribal life towards more 'civilized' values and patterns of behaviour.

Darwin ventured on to thorny ground when he applied these notions to human culture and behaviour. His naturalism recast the notion of human diversity into strictly evolutionary and biological terms, reinforcing nineteenth-century beliefs in racial superiority, where whites rested comfortably at the top of the scale. He also revealed that he believed in innate male superiority, honed by aeons of hunting and fighting. Whereas he felt that much of the animal kingdom was governed by female choice – that female birds choose their mate according to display, song or nestbuilding behaviour – he regarded advanced human society as patriarchal. In civilized regimes he felt it was self-evident that men, because of their well-developed intellectual and entrepreneurial capacities, ruled the social order and that they would do the choosing. In this way he applied biology to human culture and saw in every society a 'natural' basis for male-centred behaviour. After publication, early feminists and suffragettes bitterly attacked this doctrine, feeling that women were being 'naturalized' into a purely biological, submissive role. Many medical writers understood Darwin to be supporting the assumption that women's brains were smaller and less evolutionarily developed than men's, or that the female body was especially prone to disorders if the reproductive functions were denied.

The rest of the book tackled hot topics such as the development of human morals from animal emotions and the onset of speech (which drew Darwin into more debate with

Friedrich Max Müller). Darwin needed to show that language was not a fundamental dividing line between mankind and animals. Unlike Müller, Darwin thought that speech emerged from imitating natural sounds. 'It does not appear altogether incredible, that some unusually wise apelike animal should have thought of imitating the growl of a beast of prey, so as to indicate to his fellow monkeys the nature of the expected danger. And this would have been a first step in the formation of a language.'[8] Darwin was daring when dealing with the religious sense, proposing that this was ultimately nothing more than a primitive urge to bestow a cause on inexplicable natural events.

He also discussed likely fossil intermediaries between ape and human and mapped out a provisional family tree, in which he took information mostly from fellow evolutionists like Haeckel and Huxley. Even though there were by then a few isolated fragments of Neanderthal skulls available for study in European museums, these had not yet been identified as from ancestral humans. Huxley, for instance, regarded the original puzzling fragment from the Neander river valley as part of the thickened skull of a congenital idiot. The real advances in understanding fossil mankind were to take place several decades after Darwin's death. Nevertheless, Darwin put forward a proposal. At some point, Darwin suggested, anthropoid apes descended from the trees, started walking erect, began using their hands to hold or hunt, and developed their brains.

> The early progenitors of man were no doubt once covered with hair, both sexes having beards; their ears were pointed and capable of movement; and their bodies were provided with a tail… our progenitors, no doubt, were arboreal in their habits, frequenting some warm, forest-clad spot. The males were provided with great canine teeth, which served them as formidable weapons.[9]

The book closed with a flourish. At the end, he echoed Huxley's battle with the Bishop of Oxford at the British Association some ten years previously by saying that he would rather be descended from a brave little monkey than from a savage who delights in torturing his enemies.

Scholars nowadays agree that *The Descent of Man* offered a far-reaching naturalistic account of human evolution but did not change many minds. The people who already accepted evolution continued to believe. Those who did not accept evolution continued to disbelieve. Few readers wished to shrink the gap between mankind and animals quite so dramatically however. If these ideas were accepted, wrote the *Edinburgh Review*, the constitution of society would be destroyed. Wallace was generous about the book, praising it in letters and in reviews. Most reviewers noted Darwin's evident sincerity and depth of learning. Nevertheless there must have been a sense of déjà vu. The animal–human boundary, the human soul and the divine origin of human morals had been the main topics of debate for ten or twelve years. Young rationalist thinkers like Leslie Stephen spoke for many of the coming generation by saying 'What possible difference can it make whether I am sprung from an ape or an angel?'

With *The Descent of Man* and *The Expression of the Emotions in Man and Animals,* which appeared a year later, Darwin completed the account of evolution he began with *Origin of Species.* None of his other later writings had anything like the same public effect, although several of his final pieces, for instance on mental processes in infants, stimulated researchers. His last book was on earthworms (1881), one of the most popular books he ever published, full of natural history observations made on worms from his own garden, a symbolic and peaceful occupation that provided him with much pleasure in his fading years. Towards the end he slowed down, preferred to work on plants, and be with his family. In his seventies he enjoyed writing a little autobiography, not intending it for publication. In it he reviewed his life with great charm and modesty. Yet what a life it had been. Few men reach such heights of intellectual power or have their views discussed so widely and with such vigour. Even if people did not think that they were descended from apes, they talked about it ceaselessly.

For those who did believe, Darwin became a kind of prophet, a secular saint. From the middle of the 1870s his life took on many of the trappings of celebrity culture, rather as Charles Dickens, opera singer Jenny Lind or other famous figures in the Victorian period discovered to their cost. Darwin's portrait was circulated in illustrated magazines, he received requests for autographs, free copies of his books, money and advice, and his home was visited by sightseers, keen to catch a glimpse of a man whose work had so notably contributed to nineteenth-century debate. The years of controversy generated extraordinary fame. Young scientists increasingly asked to be admitted to his presence for a kind of personal benediction, either to eat lunch with the family, or to enter his study, which became in people's minds an inner sanctum, the place where great thought had taken place.

Loved by his family, appreciated and admired by his friends, an intellectual beacon to many, in turn respected and reviled, Darwin came to the end of his life knowing that he had brought about an extraordinary transformation in scientific thought. His identity had become subsumed in that of his book. 'If I had been a friend of myself, I should have hated me,' he remarked with some humour to Huxley at the height of the controversy. 'I wish I could feel all was deserved by me.'

CHAPTER 5

Legacy

Twenty-three years after publishing the book that made him famous, Darwin died at home, aged seventy-three. He was buried in Westminster Abbey, in London, the more usual location for state funerals, royal marriages and national celebrations. Such a burial site for the author of *On the Origin of Species* was ironic in many ways, for the nation was well aware of Darwin's reputation for having undermined church authority. By the time of his death, however, Darwin was fêted as a great scientific celebrity, a grand old man of science, someone who had looked further and seen more than others, of an intellectual rank as great as Newton, and certainly deserving to be honoured in the country's primary commemorative setting. Professors, churchmen, politicians, medical luminaries, aristocrats and members of the public crowded the Abbey to see him to the grave. 'Happy is the man that findeth wisdom' sang the choir. It is hardly possible nowadays for us to guess whether Darwin died a happy man but he was certainly revered for his achievement and personal character, the very model of what a man of science ought to be.

However, despite this reverence, the cultural world was entering a different phase, recognizably more modern in tone. The fierce religious controversies of earlier days were subsiding. By regarding the Bible as an allegorical text filled with spiritual meaning, it became possible for Christian believers to retain their belief in the truth of God's message while also appreciating scientific findings as a different kind of truth. Moreover, the power of the Church itself was on the wane. Many of these changes were retrospectively attributed to the *Origin of Species*. Honours paid to Darwin at his funeral liberally acknowledged his important role in constructing the modern frame of mind.

His scientific legacy, though, was not nearly as secure. As fresh areas of research opened up in the biological sciences, and new kinds of professionals took up a wider range of problems with more sophisticated techniques, the original thesis of natural selection was modified almost beyond recognition. There was dispute about the central concepts of competition, success and 'fitness', particularly in the way these interlaced with contemporary political ideologies. Alternative evolutionary systems based on direct responses to the environment came into play. Indeed, it is frequently said that Darwinism was eclipsed by other systems of evolutionary thought towards the end of the nineteenth century, not to be restored until a 'new synthesis' was put forward in the 1940s.

Much of this eclipse rested on fresh critiques of the main struts of Darwin's original proposals. Social Darwinism was criticized as it climbed to pre-eminence in political thought around 1900. Wallace came to reject the competitive aspects of Darwinian biology as applied to human society and supported utopian socialist principles. Elsewhere J. Keir Hardy argued that progress took place via group selection in which individuals felt sympathy for one another. In Russia, the prevailing ideology was that the main struggle for existence was not species against species, but species against the environment. The émigré Russian Prince, Peter Kropotkin, pushed this furthest in *Mutual Aid* (1902), arguing that evolution's main driving force was cooperation, exactly the reverse of competition. Socialist thinkers such as George Bernard Shaw insisted on the moral superiority of Lamarckian ideas, where the effects of the environment were believed to be more important in shaping human character than inbuilt biological properties. J. B. S. Haldane confidently declared 'Darwinism is dead'.

The operating mechanism of selection was criticized too, encouraged by the work of the young critic and writer Samuel Butler (1835–1902). Butler's *Evolution Old and New* (1879) downplayed Charles Darwin's scheme in favour of those of Dr Erasmus Darwin and Lamarck. Butler proposed that Charles Darwin was merely one in a long line of evolutionary thinkers, and that *Origin of Species* misdirected biologists to seek struggle and mechanistic answers where older schemes had far more to offer in recognizing that organisms might respond adaptively to the environment. Butler and Darwin had argued fiercely in the last years of

Darwin's life over the text of a biography of Dr Erasmus Darwin, a quarrel that had begun in an unfortunate breach of etiquette on Darwin's part and quickly came to represent a clash between generations and world systems – for Darwin was unable to control Butler as he was accustomed to control his other disciples. It ended in complete personal estrangement. This quarrel intrigues historians because of the way it reveals cracks opening up in the Darwinian edifice. Butler's views chimed neatly with increasing debate over the relative roles of heredity and environment, not only in biological theory but also in understanding human mental development from child to adult and the structure of society. Galton's catchphrase of 'nature or nurture' (biology or environment) became an issue of considerable concern.

Furthermore, even though there was great enthusiasm among naturalists for reconstructing the history of life on earth, it soon appeared to be the case that non-Darwinian, pre-directed paths of evolution were more attractive. Palaeontologists took a lead in this area, probably because of the spectacular fossil discoveries of the late nineteenth and early twentieth centuries in the American West. The American palaeontologist Theodore Eimer claimed that evolutionary history had not taken the shape of a Darwinian branching tree but proceeded in a straight line. In his eyes, natural selection was powerless except to weed out obviously deleterious trends. A much-discussed example was that of the Irish Elk, which was thought to have become extinct because of the dramatic over-development of its antlers – the suggestion was that the antlers had acquired a momentum of their own and eventually became a liability not an advantage.

Alpheus Hyatt, another noted fossil expert, similarly argued that adaptive trends almost always carried on beyond their usefulness. Ultimately, he said, a species would be driven to 'racial senility' and extinction. His colleague, Edward Drinker Cope, alternatively felt that evolution roughly followed the same course as the embryo of an individual, sometimes accelerating, sometimes dropping back. Henry Fairfield Osborn, the director of the American Museum of Natural History in New York, one of the world's greatest natural history museums, and a committed Darwinist, believed that each group of organisms experienced a period of rapid diversification at the start of its history, which then stabilized into several steady lines of development. Like Eimer and Cope he did not see in the fossil record any of the multiple branching described by Darwin. Indeed, he claimed entirely different animal groups might progress along roughly the same routes, in the development of horns for example.

Such straight-line evolutionary histories, with their subtexts of inbuilt senescence or death from over-specialization, lent authoritative support to increasingly pessimistic views about the human future. Primitive cultures could now be regarded as in the 'infancy' of their development. More advanced societies might be set on lines of development that led them through the heights of civilization to corruption or decay. Those who transgressed society's conventions, such as criminals, homosexuals or the mentally deranged, could be categorized as 'throwbacks' to some racial past. As optimism in continued progress drained away, such concerns were vividly expressed in late nineteenth-century fiction. H. G. Wells's *Time Machine* (1895) took a traveller into a future where humans had deteriorated into two species, the brutal underground Morlocks and the effete overground Eloi, a parable of the political and social divisions that Wells discerned in his own day. Bulwer Lytton's *The Coming Race* (1871), Samuel Butler's *Erewhon* (1872) and Arthur Conan Doyle's *The Lost World* (1912) played on largely the same themes, while Emile Zola and Thomas Hardy drew powerfully on the idea of hereditary degeneration and the inflexible pull of biological forces on humanity.

By the start of the twentieth century much of the developed world was caught up in eugenic and hereditarian systems of thought on a wider scale. Eugenic movements reached a peak in 1912 with the first International Eugenic Congress, held in London. Long before then, Francis Galton and others in Britain caught the pessimistic mood of the times and pointed to the poor quality of army recruits for the Boer War to illustrate the decline of the nation's biological fitness. Other signs of 'degeneration' seemed to abound in the eyes of the elite: increased criminal behaviour, a loosening of moral values with a consequent rise in prostitution and venereal diseases, a growing political restlessness among the workers, unionization and the threat of strikes or demonstrations. Huge publicity surrounded the legal case brought against Oscar Wilde for homosexuality. Even the cause of women's suffrage and the political prominence of the 'new woman' (women who worked, who wished to be educated and to vote, and perhaps bicycled and smoked cigarettes) were taken as symptoms of a nation in decay. Whereas in Darwin's day eugenics was mainly expressed in fears about the maintenance of biological fitness, in the early twentieth century it expanded through Europe and the Americas into significant political movements seeking to change government policy with public health measures for the masses, birth control and enforced restraint from breeding. At root, the old system of Malthusian checks that Darwin had applied to biology was reapplied to political economies with compelling biological support. The poor, the deranged, the weak and diseased came to be regarded as biological burdens on society. For the good of the nation, it was said, policies should be introduced to prevent them from reproducing their kind.

Many of these initiatives took an institutional form. A National Eugenics Laboratory was established in University College London with a bequest from Galton to investigate deteriorating family lines, principally gauged by the incidence of hereditary mental disorders. It was headed by Karl Pearson, an idealistic eugenicist and Darwinian biologist with marked socialist leanings. Psychiatrists identified degenerative 'types' among their inmates using the new medium of photography, and criminologists such as the Italian writer Cesare Lombroso proposed that there were physical stigmata to be seen in social deviants. These were sometimes linked explicitly with apish bodily features. He also popularized the word 'atavism', meaning a reversion to some ape-like ancestral type. Conditions such as epilepsy or gross deformity categorized yet others again as undesirable. It was thought that such unfit individuals could be identified by 'signs' and then removed from society. In 1888 the Parisian detective Alphonse Bertillon did exactly this by introducing a system of physical signs and measurements to identify any individual who came through the French criminal system, including the technique of taking fingerprints, the basis of all modern identification procedures. The same threat of physical and moral degeneration was taken up in dazzling fashion by Robert Louis Stevenson's *The Strange Case of Dr Jekyll and Mr Hyde* (1886) where Jekyll's other self, the evil Hyde, progressively became more apelike as his murderous deeds increased.

Urban decay, industrial squalor and a wish for interventionist public health measures such as vaccination and the regulation of prostitution, filled the public journals. Upper-class fears in Britain about being overrun by a depraved and criminal underclass (the 'mob') became widespread. The Eugenics Education Society, soon to be the Eugenics Society, was established in Britain from 1907, and rapidly filled up with earnest professional people wishing to improve and control the masses. Its president from 1911 to 1925 was Leonard Darwin, one of Charles Darwin's sons. An important outcome in Britain was that the Mental Deficiency Act was passed in 1913 to identify mentally impaired individuals and segregate them in an institution or asylum where they would be prevented from reproducing. Other European governments, particularly in Scandinavia, moved decisively in the same area, although some of these laws were never put into practical operation. All too often it turned out that the poorer sections of society contained the larger proportion of unfit individuals. Procedures were peremptory. Asylums, orphanages and prisons became dustbins for undesirables.

In America, too, eugenics flourished in the early twentieth century. In 1910 the Eugenics Record Office at Cold Spring Harbor was founded and efforts were made to trace traits such as insanity, feeble-mindedness and criminality back through the generations. The first task was to identify those who should not reproduce. Hereditary forms of mental disorder became the main target. Among the most notorious eugenicists, Dr Henry H. Goddard, of Vineland, New Jersey, adopted the French system of intelligence testing to compute the mental age and ability of mentally defective children, which were quickly converted into tests for IQ (intelligence quotient). Goddard coined the terms 'feeble-minded', 'imbecile' and 'moron' to describe specific levels of impairment and proposed that such people should be permanently separated from the rest of the population. He did not carry out sterilizations, although some medical bodies recommended that this should take place. He did, however, provide the government with a quantitative framework – a test – for identifying the biologically unfit in society. Later, Robert Yerkes tested the adult male population called up for service in the First World War (some 19,000 servicemen). He calculated that most of them possessed a mental age of thirteen years old. His IQ tests further indicated that African-Americans and others of recent European origin had even lower mental ages. Prostitutes and the Polish were lowest of all.

These tests were evidently biased in favour of literate middle-class whites familiar with North American culture, a fact made further apparent on Ellis Island. Tired, traumatized and usually unable to speak colloquial English, many hopeful immigrants to the USA were incorrectly categorized as imbeciles and turned away. Goddard's statistics deeply shocked the American government. Charles Davenport, the director of the Eugenics Record Office, advocated the introduction of state programmes to restrict marriage, enforce segregation and compulsory sterilization. During the period 1900 to 1935, no fewer than thirty-two states passed sterilization laws. Most of the 60,000 people known to have been sterilized under these regulations were mental asylum patients or prisoners. It is not recorded how many were of African descent.

Eugenic doctrines around 1900 were invariably coupled with other ideological extensions of Darwinism. Several biologists and eugenicists working within the Darwinian system threw their support behind Germany's claim to be Europe's leading nation, particularly Haeckel who proposed a materialist philosophy of life called 'monism' in which spirit and matter were different aspects of the same underlying substance. His Monist League promoted German supremacy in the decade before the First World War and indirectly contributed to the rise of fascism afterwards. Embedded in these biologized aspects of society and visions of national ascendancy, Germany's rulers reached furthest of all with their eugenic law for the Prevention of Genetically Diseased Progeny (1933). Some 300,000 people were sterilized under this edict until 1939 when it was replaced by the wartime 'euthanasia' programme for the extermination of the Jews.

Race science, sometimes known as racial science, reflected the most extreme prejudices of the day and this too drew on Darwinism. It should be said, however, that racism and genocide predated Darwin. Nor were they solely confined to the West. Nevertheless, evolutionary views, and then the new science of genetics, gave powerful biological backing to those who wished to partition society according to ethnic difference or promote white supremacy. The American author Joseph Le Conte spoke for many when he justified the subjugation of blacks in the post-Civil War South by saying that 'the negro race is still in childhood... it has not yet learned to walk alone in the paths of civilization'. Some racial scientists believed that different ethnic groupings were completely separate species, although this was always a minority view. Carl Vogt's theory, for example, was that each race had evolved from a different ape: whites from the chimpanzee, blacks from the gorilla, and orientals from the orang-utan. In Europe and North America these and other racial scientists debated human interbreeding, made prurient ethnological investigation into sexual behaviour and initiated studies of mixed breeding in former slave-owning regions. Universities and museums accumulated collections of skulls from all over the world for scientists to measure cranium capacity (thought to be an indicator of intelligence) and deviation from a supposed ideal Caucasian type. These collections, a relic of long-superseded theories, are now an embarrassment to national institutions and are never put on display.

Armed with the naturalist Gregor Mendel's notions about the transmission of characteristics from one generation to the next, a fresh generation of theorists turned the study of human evolution into a science of racial fixity that legitimized contemporary prejudices.

For Americans, the race question not only highlighted the problems created by slavery and difficulties encountered with emancipation after the Civil War but also precipitated academic warfare between social scientists and biologists, the former favouring cultural explanations of racial differences, the latter inbuilt physical and biological parameters. Franz Boas, one of the founders of anthropology, who argued for the unique and equal nature of every culture, suffered at the hands of a powerful race lobby within American biology during the 1920s that endorsed the existence of stages through which every society must pass in its development. Across the Atlantic, at much the same time, the Nazis claimed that the Aryan was a distinct and superior form of humanity destined to rule over 'sub-humans'. Subsequent horror at the Nazis' drive to eliminate the Jews challenged the ideology of racial science, although much still exists.

Racial theory of a lesser kind was also put to use by early twentieth-century palaeo-anthropologists who started to suggest that there had been multiple lines of human evolution, with some of those lines, including the Neanderthals, being driven to extinction by more successful races at various stages in the process. As traces of fossil humans began to emerge, scholars became convinced that there must have been a series of intermediary animal-man forms. In retrospect it is intriguing to see how much naturalists wanted to make these intermediaries apelike in shape and character, especially insisting on the small size of their braincase. The so-called 'human' characteristics were thought to appear quite recently in geological history, almost all together in a rush with the emergence of *Homo sapiens*. Eugene Dubois was famed for his exciting discovery of 'Java Man' in 1891, an ape-man that he named *Pithecanthropus*. The discovery of another species, to be named 'Pekin Man', arrived in the 1920s. Raymond Dart's 'Taung baby' added a South African species named *Australopithecus*. The prize of becoming known as the cradle of mankind generated bitter national rivalries for fifty years or more. Soon, an exhibit in the American Museum of Natural History displayed reconstructions of three types of extinct mankind, Pithecanthropus, Neanderthal and Cro-Magnon (very close to present-day mankind), arranged as a progressive series towards the white, civilized form of today.

This fascination with ape-men perhaps accounts for the ease with which a notorious fraud was accepted by the academic community. The remains of an early human skull and jaw were found by an amateur archaeologist, Charles Dawson, in a quarry near Piltdown in East Sussex in 1912, and described as a new species of intermediate hominid, *Eoanthropus dawsonii* (Dawson's dawn man). These bones fitted well with the hypothetical line of human evolution then in favour. Sir Arthur Keith, for example, one of the leading investigators of early mankind in Britain, regarded the remains as from a lower type, with no close relationship to the other humans known to have existed at the same time. Keith had little doubt about what happened when higher and lower forms came into contact: warfare between the races was a natural part of the prehistory of human evolution, he confidently declared, in the same way as the terrible 1914–18 warfare of his own time had resulted in victory for the British, the survival of the fittest. Gradually, however, the Piltdown skull came to be seen as increasingly anomalous. In the 1950s it was exposed as a hoax: an ape jaw had been attached to an ancient human skull and the teeth filed down to produce a human pattern. Dawson was probably not the main culprit. Other maverick amateurs have been suggested, each one with a grievance against the scientific establishment.

Across the globe fundamental shifts were taking place in the way scholars thought about the natural world. Modernism was on its way. Growing numbers of biological scientists started turning away from the problem of how species existed in the wild or the history of the evolutionary tree to direct their attention down into the living body, seeking the mechanisms of inheritance, hybridization, mutation and variation. At the time of Darwin's death many already believed that inheritance held the key to life. By the last decade of the nineteenth century their aim was not to catalogue dead animals and plants but to understand the inner workings of living, breathing bodies – a self-conscious conceptual break with the past. This new attitude to biology reflected a major move away from observational natural history towards a more experimental, laboratory-based form of investigation, a move that can be seen taking place in almost all of the sciences at this time. Traditional natural history, of course, did not stop; it became sidelined, sometimes regarded as the province of amateur naturalists, or otherwise reconstituted as new sciences of animal behaviour, ecology and environmentalism. Like physics and chemistry, biology was becoming something that was primarily practised indoors, in a lab, under controlled conditions, and increasingly with the financial aid of government agencies.

These new experimental biologists made many astonishing discoveries in a relatively short period of time. Some pressed deep into the building blocks of the living body, investigating the cell and early stages of embryonic development. Others explored remaining gaps in Darwin's theories by studying variation and inheritance. Galton's speculations about innate inherited traits seemed to answer some of

Darwin's unanswered questions. But Galton's proposals were entirely abstract ideas, never quite realizable in a laboratory setting. A group of his followers, clustered around Karl Pearson in the eugenics laboratory at University College London, therefore began to study how inheritance and variation might work in practice. Calling themselves biometricians, these men (and a few pioneer female scientists) measured variability in living beings, for example the dimensions of crab shells, and devised many of today's most common statistical procedures for calculating deviations from the norm in order to show small adaptive shifts in a chosen species. By 1900 they were perhaps the last truly committed Darwinians in existence, for they insisted on Darwin's original system of slow, gradual changes in populations.

In other parts of life, biometricians were quick on the draw. For five years or more, they quarrelled violently with a rival group of biologists at Cambridge University under the eye of William Bateson (1861–1926), himself an excellent field naturalist and experimental hybridizer. The Cambridge group was adamant that evolution proceeded by jumps and starts, and that columns of statistics produced in London were not going to tell anyone anything about how animals and plants varied or transmitted their characteristics to offspring.

This controversy has often been understood as the foundation of modern genetics for it provided the context in which Mendel's work on peas was rediscovered. Three noted European experimentalists, Hugo de Vries, Carl Correns and

Erich von Tschermak, each independently working on the variation of plants, and individually keen to disprove the biometricians' arguments, one by one encountered Mendel's paper in the early months of 1900 and brought it to public attention. As they put it, the essence of Mendel's experiments was to show that the heritable characteristics were self-contained and not able to blend – in Mendel's research, the peas in the pod were green or yellow, smooth or wrinkled, never anything in between. These self-contained characteristics tended to reassort (rearrange) themselves during the reproductive process and appeared in fixed proportions in subsequent generations, say three wrinkled peas for every smooth one. Moreover, the characteristics could be either dominant or recessive: that is, some were visible in the body of the offspring while others remained hidden. Mendel had no notion of the modern 'gene' and yet his work strikingly anticipated the key concept of twentieth-century genetics that most physical characteristics, every pair of brown eyes, could be linked to a single particulate entity that was sorted and transmitted independently from generation to generation.

Nor could Mendel have anticipated how his results would be used. Bateson enthusiastically appropriated Mendel's findings, turning his group at Cambridge into the first Mendelians in the world. Their approach was decidedly non-Darwinian, in the sense that they believed Mendel's results supported the idea that evolution operated by jumps based on relatively sudden variations or 'mutations' in organisms. To them, the continuous tiny changes stipulated by Darwin and so carefully measured by the London biometricians were irrelevant, a waste of good scientific time. Within a few months, the transformation was complete. Bateson named his new science 'genetics' – the study of heredity – and claimed that mutation theory supplied the answer to the origin of new species.

Indeed, the science of genetics at the start was somewhat anti-Darwinian. For more than twenty years its practitioners proposed that mutations were the source of new and favourable kinds of organisms – happy accidents that would introduce an entirely different kind of being into the natural world. These early geneticists had no need for natural selection. It took a lot of dedicated work in the 1930s and 1940s to see how Mendelism and Darwinism might be brought together.

Meanwhile, close attention was paid to identifying the inheritable material and how it was transferred from generation to generation. At that time, it was not at all obvious how chromosomes might be involved. In 1893 August Weismann proposed that there must be an invisible substance that carried all the hereditary information from parent to child. He called this 'germ plasm' and claimed it could not be affected by the environment. This germ plasm played a useful interpretative role until expanded by Wilhelm Johannsen's definition of the 'gene' in 1911. Even Johannsen was unsure if the gene really existed until Thomas Hunt Morgan, the outstanding geneticist of Columbia University, New York, demonstrated that genes were, so to speak, real entities strung along the chromosomes like pearls on a necklace and that they definitely contained the heritable material. Morgan's famous experiments depended on one particular experimental organism, the fruit fly *Drosophila melanogaster*, which happened to have large and easily visible chromosomes. By breaking or otherwise manipulating the chromosomes, Morgan's laboratory team produced a succession of mutant flies, for instance with red eyes or fused wing covers. The work proceeded with such sophistication that the team could locate which part of the chromosome was specific to each mutation. The results were summarized in *The Mechanism of Mendelian Heredity* (1915), now regarded as a milestone in modern genetics, and for which Morgan received the Nobel Prize. His book was completely non-Darwinian. With a fine new theory of chromosomal mutations and the gene to answer every question, Morgan discarded Darwin's ideas of variation, adaptation and selection.

The influence of the *Origin of Species* was subsiding elsewhere too. Other geneticists favoured environmentalist ideas of inheritance. Soviet communist governments were generally hostile to the capitalist implications of Darwinian theory in the twentieth century and endorsed a revised form of environmentalism brought into state policy by Trofim Lysenko in the course of the 1930s. Lysenko's achievement had been to demonstrate the adaptation of wheat to prevailing climate conditions ('vernalization', in which the seeds were exposed to cold so that they would germinate earlier the following year). Lysenko claimed this property could be inherited and thus new breeds of wheat could be produced suited to the short growing season in Russia. Stalin adopted Lysenko's findings, forbade alternative genetic research and instigated a purge of leading geneticists, notably Sergei Chetverikov and Nikolai Vavilov. Some fled to the West, such as N. W. Timoffeef-Ressovsky and Theodore Dobzhansky, and there contributed extensively to the rise of new genetic ways of thought. Others simply disappeared. Under this regime, reports of amazing (and impossible) agricultural successes were issued until the middle years of Khrushchev's power, when Lysenko was denounced by the physicist Andrei Sakharov. It was the mid-1960s before Russian science gradually opened up to Darwinian ideas of evolution and the new genetics.

By the 1930s, in fact, it was difficult to see exactly where Darwin's theory might still be relevant. Molecular biology was beginning; chemistry and physics were increasingly used to explore the inner structure of living matter; and laboratory techniques were making substantial advances in understanding the workings of the cell and mapping the genetic basis of heredity. Field naturalists felt themselves to be left behind in the academic contests of big biology. From today's perspective it is almost unimaginable to envisage a world of biological research without the concepts of adaptation and natural selection, the intellectual tools that underpin so much of modern biomedicine, the environmental sciences, theories of human behaviour and psychology. So what might have stimulated a mass revival of Darwinism in the middle years of the century?

Historians agree that three diverging lines of research were forcibly brought together by a group of inspired young naturalists in the 1940s, a group including the writer and biologist Julian Huxley (Thomas Henry Huxley's grandson), Ernst Mayr, an émigré field naturalist and philosopher-biologist from Germany, Sewell Wright, the American geneticist, George Gaylord Simpson, a vertebrate palaeontologist and G. Ledyard Stebbins, an up-and-coming botanist and geneticist. The story of twentieth-century Darwinism lies with these figures who struggled to give it new meaning and integrate it with cutting-edge experimental disciplines. Putting the situation somewhat starkly, the field and observational naturalists who continued to feel themselves directly connected to Darwin's own work had to reinvent themselves. Although they scarcely intended this to coincide with any other event, the 'modern synthesis' was in place just in time for lavish centenary celebrations of the publication of the *Origin of Species* in Chicago in 1959.

An important first step was the reconciliation of Darwin's original proposals with early twentieth-century genetics. In effect, it was necessary to turn the external process of animal and plant evolution into changes in the frequencies of genes. Repeated small mutations in the chromosomes were consequently reinterpreted as building up the fund of variability needed for the raw material of selection. Every trait, it was now realized, exhibited a continuous range of variation, so that in a large population there would be plenty of differences circulating through the gene pool on which selection could work. One of the leading figures in this movement was the Cambridge statistician Ronald Aylmer Fisher, who created a mathematical model to show how the frequency of a favourable gene could increase in a population. Fisher devoted a significant portion of his resulting textbook to discussing the human implications: inspired by Pearson he was an ardent eugenicist as well as a liberal Christian who claimed to see God's hand in biological progress. Another significant figure was J. B. S. Haldane, a larger-than-life individual who contributed notably to British public education in the interwar period. Like others at the time, Haldane looked enthusiastically to Marxism. He campaigned against Fisher and eugenics. Haldane ultimately resigned his professorship at University College London in protest at Second World War militarism and went to teach in India.

The man who turned it all into a theory of population genetics was Sewell Wright at the University of Chicago. By 1920 Wright had developed a powerful mathematical procedure to explore the flow of genes in small populations of laboratory guinea pigs and hooded rats. He investigated natural populations in the 1930s and proposed that similarly small groups in the wild must be subject to what he called 'genetic drift'. Wright's metaphors of an adaptive landscape with mountain peaks and valleys proved an effective way of thinking about the extension or contraction of small knots of particular variations inside a larger population, each little group ready to rise or fall in numbers according to changing conditions. Wright's work was made more widely available through successive editions of Theodore Dobzhansky's landmark textbook, *Genetics and the Origin of Species* (1937).

Inspired by the fresh ideas in population biology, Ernst Mayr settled down at Harvard University to integrate his ornithological field studies with genetics. Of all the biological thinkers of the twentieth century, Mayr was perhaps unique in his grasp of both practical detail and philosophical vision. Like Darwin, he concluded that a new species might develop if a variant group of organisms was geographically isolated in some way from its parent population. Dobzhansky added to Mayr's proposals by suggesting that there were probably other isolating mechanisms as well, such as behavioural characteristics or different breeding times, all of which would prevent two or more populations from merging. At the same time, G. G. Simpson reinterpreted the fossil record, smoothing out its stops and starts to accommodate the idea of continuous variation. He argued that transitional forms would be rare and therefore infrequently preserved, giving to the fossil record a false appearance of sudden big changes. Then Stebbins showed how a plant's occasional doubling and trebling of the chromosomes could explain the sudden origin of dramatically different species in the plant world. All three managed to unite the apparent discontinuities of the living world with a genetically-aware reinterpretation of Darwin's small and gradual steps. Julian Huxley brought them all together in a popular book published in 1942 called *Evolution: The Modern Synthesis*.

The only thing this group did not have was proof. Ever

resourceful, and in much the same way as Morgan had found a Nobel prize in fruit flies, the newly reconstituted Darwinians turned with delight to the Galápagos finches and then the peppered moth, *Biston betularia*. The Galápagos finches subsequently became the best-known example of evolution in the world – not through renewed attention to Darwin's writings, it must be said, but through the work of David Lack, a schoolteacher and amateur ornithologist.

Lack had come to Julian Huxley's attention in 1938 and visited the Galápagos Islands soon afterwards to observe finch behaviour for an entire breeding season. After ten years' further work in museums, he concluded that the beaks held the key to their evolution. Each species had become adapted to a particular foodstuff, thereby allowing diversification into many different niches. His book, *Darwin's Finches* (1947), described the birds as an example of evolution in action. Featuring in countless biology textbooks, nature documentaries and popular evolutionary accounts, 'Darwin's finches' rapidly became synonymous with the new Darwinism. The work carried out by Peter and Rosemary Grant from the 1970s in the Galápagos Research Station still provides the most influential field study of evolution ever conducted.

The peppered moth was equally successful. It became an iconic example of natural selection just in time for the 1959 centenary celebrations of the *Origin of Species*, although the case afterwards became surrounded by unsubstantiated accusations of fraud. The study was made in Britain by

Bernard Kettlewell under the guidance of the pioneer Oxford University population biologist Henry Ford. The moth itself could hardly have seemed a better demonstration organism. In nature it exists in two forms, one a speckled black and white, the other a black mutation, called melanic. On ordinary oak trees, the first form is almost invisible. This advantage is reversed in polluted industrial areas where the black form is better camouflaged on darkened tree trunks. Kettlewell released quantities of both sorts of moth in two wooded sites, one near Manchester, where the trees were blackened by soot, the other in clean countryside in Dorset. He demonstrated that birds ate the most visible form, thereby operating a Darwinian selective pressure that allowed one kind of moth to survive and increase in number at the expense of the other. It categorically showed that selection could alter the frequency of particular genes (in this case the melanic gene) in a population. One summer, the famous animal behaviourist Niko Tinbergen spent a few days with Kettlewell filming wild birds picking the moths from a tree trunk. Now a natural history film classic, the old black and white film was shown on early television screens, a perfect way to display black and white moths against their black and white backgrounds. In recent years, government-controlled reductions in pollution have reduced the black form in Britain to the point that biologists now find it difficult to repeat Kettlewell's observations.

A huge step in the unification of the biological sciences had been achieved. The modern synthesis transformed the old notions of selection and adaptive change and breathed fresh life into Darwin's ideas. Biologists also took a new interest in Darwinian themes that emphasized observation and practical field studies. Many biologists at this interesting time looked back directly to Darwin himself. The centenary of publication of the *Origin of Species*, which was coincidentally also the 150th anniversary of Darwin's birth, was an occasion for much celebration and revivalism amidst the rhetoric of future scientific advance. Some biologists wrote biographies of Darwin, others edited for publication his *Beagle* journals and notebooks, and others again made moves to preserve his house as a memorial and museum in which the significance of modern evolutionary science could be appreciated and explained. In their eyes, evolutionary biology was at last a recognizable scientific discipline. Darwin was elevated into its founding father.

The new generation of Darwinians also addressed the question of human ethics. Most of them were convinced that science confirmed the absence of any underlying plan or divine purpose built into the universe. G. G. Simpson, one of the architects of the modern synthesis, pointed out that it was impossible to regard the human species as the predetermined goal of random shifts in gene frequencies. Amusingly, he said that mankind was the result of a process that never had him in mind. The modern synthesis, in fact, was much less readily compatible with spiritual belief than any previous, more flexible, theistic evolutionary theory. From the 1950s, there was an increasing tendency for practising scientists to be disbelievers, at the very least when inside their labs. The essence of modern science, it was commonly said, was to seek answers in the world of evidence and proof, not to call on the divine or other supernatural factors.

A few found spiritual consolation in continued ideas of social progress. Scientific naturalism could take on the mantle of a religion, as once preached by Thomas Henry Huxley in his 'lay sermons' or articulated by philosophers William James and Charles S. Pierce. The evolutionary mysticism of Pierre Teilhard de Chardin's *Phenomenon of Man* (1959) was also popular among those seeking spiritual guidance in evolutionary processes. The world of living beings, he suggested, was enclosed by a sphere of mental unity, called 'Noosphere'. This predated the idea of cyberspace by some twenty years, and Teilhard de Chardin is primarily remembered now for influencing the engineers of silicon valley. Julian Huxley was generally sympathetic to these views and promoted a philosophy of humanism, rejecting the idea of a transcendental creator, but drawing on older nineteenth-century idealism to stress the human race's responsibility to foster moral progress. As it happened, most biologists were willing to believe that humans were still special. Human intelligence, adaptability and social characteristics were still seen as indications of a higher level of development than in animals. Humanists felt that the human race was capable of moving onwards from biology to make a better world based on pacifist, altruistic social policies.

After the brutality of the Second World War, however,

there seemed little point in glossing over the harsher side of animal behaviour. The founder of modern ethology, Konrad Lorenz, demonstrated the innate aggressive behaviour of animals and warned that humans, too, were endowed with similarly destructive basic instincts. The message was reiterated by Robert Ardrey in his work on the 'territorial imperative' and by Desmond Morris in a widely distributed popular text, *The Naked Ape* (1967). Soon the terminology of primate studies was spreading from science into common public usage. Advertising moguls enjoyed a particularly inventive time with their slogans about 'alpha-males'. To be human, it seemed, was to be brutish.

Such an image of human nature as fundamentally selfish and aggressive did not go unchallenged during the peace demonstrations and love-ins of the 1960s. The grand old man of palaeo-anthropology, Louis Leakey, encouraged three female scientists to pursue actual observations of apes in the wild, the first time that this had ever been achieved to modern scientific standards. He placed Jane Goodall in Gombe Stream Reserve, near Lake Tanganyika in East Africa, to observe chimpanzees, and Biruté Galdikas in Sumatra for the orang-utans. Last, he put Dian Fossey to work in a gorilla reserve in Rwanda from 1967. These studies of apes in their natural habitat showed them generally to be family-oriented, loyal to their troop, and not aggressive unless frightened. As a result, there was renewed willingness to take seriously a closer mental and emotional relationship between apes and mankind. With wide connections to the public through magazines like *National Geographic*, these alert observational scientists were also among the first to stimulate political awareness of conservation issues.

Tension between such notions has never subsided. Discussion of the thin boundary lines between animals and mankind, between science and human values, has most recently taken its cue from Edward O. Wilson's *Sociobiology: The New Synthesis* (1975) in which animal and human behaviour patterns are located in the genetic framework of each species. Wilson argued that all organisms are genetically programmed to guarantee the most reproductive benefit to themselves: males tend naturally to spread their ample sperm around as far as possible and females to conserve their valuable eggs. Males do not stick around to look after the baby, females search for the best, most committed father. All behaviour patterns could be linked more or less back to the genes' drive to survive. In claiming this, Wilson did not deliberately intend to suggest that human lives are completely biological, although he did say that 'the gene holds culture on a leash'. Nor did he propose that humans are little more than a bundle of genes. He agreed that societies are mostly constrained by political institutions, economic limitations and social convention.

Yet to critics, this deterministic approach, rooted in an unyielding science of genes, is difficult to distinguish from dangerously ideological uses of genetics. Sociobiology could easily be used to support claims for inbuilt differences in intellectual ability, ethnicity or gender roles. Religious thinkers resent the notion that moral values apparently derive from biological utility: that a mother cares for her child in order to ensure that her genes are successfully passed to the next generation. Left-wingers fear that such ideas might be taken up by the political right to justify conventions like the nuclear family or to avoid civic improvements and medical care because it seems easier (and cheaper) to believe in unalterable, hereditary, biological traits. People from the humanities decry a continuing reduction of human attributes to mere biology. This cultural and scientific argument continues unabated through the twenty-first century.

In 1976 a widely read text by Richard Dawkins called *The Selfish Gene* brought many of these issues to the fore. Dawkins explained the world of genes metaphorically, as if every living organism, every songbird or chimpanzee, was merely the gene's way of making another gene. Behaviour patterns were little more than useful devices for ensuring the reproduction and spread of genes in a population. His lively terminology caught the imagination. Like Wilson, Dawkins has often been criticized. Whipped up by sensationalist headlines in the mass media, the public now tends to think that science proposes a gene for every human characteristic (an 'intelligence' gene, a 'homosexual' gene, an 'adulterous' gene) in the same way that there might be a gene for cystic fibrosis. Geneticists consequently find it hard to make clear that no single gene for anything ever exists, and that individual personalities or medical conditions rest in the interaction and expression of many genes via proteins in the cells and in relation to local environmental conditions, social structures and upbringing.

Few of these modern debates over gorillas, selfish genes and biologically programmed behaviour patterns, however, have generated religious controversy about the accuracy of the actual knowledge being produced. Even Pope John Paul II issued a letter in 1996 to Catholics acknowledging that the result of scientific work carried out independently all over the globe 'leads us to recognize in the theory of evolution more than a hypothesis'. The most unexpected of all recent developments, therefore, is the resurgence of creationist literature and proliferation of a whole new range of anti-Darwinian theologies in the West. Possibly it is one further expression, among many, of a cultural reaction to the relaxation of moral codes since the 1960s and 1970s. New creationists perhaps blame the rise of secular ideas for modern decadence and the loss of traditional family values. To attack evolutionary theory would therefore be to attack both a symbol and the alleged cause of the rot. Seen from outside, the tone of this movement is proscriptive and conservative. Whereas the anti-Darwinians of the nineteenth century never managed to bond in concentrated battlelines, and lost much of their effectiveness as a consequence, the fundamentalists of late twentieth-century America have acquired an impressively united voice and high public profile.

Many of these modern movements echo themes brought out at the Scopes trial in Tennessee in 1925, where politicians and theologians attempted to drive Darwinism from public education. The legislatures of six southern states had already proposed anti-evolution laws during 1923 when two lesser bills were passed. In 1925 the Tennessee House of Representatives passed a bill making it a crime 'to teach any theory that denies the story of the divine creation of man as taught in the Bible, and to teach instead that man has descended from a lower order of animal'. When the American Civil Liberties Union declared that it would defend any Tennessee schoolteacher willing to defy the law, John Scopes, a young science teacher from Dayton, accepted. The trial itself began as a publicity stunt, but soon gave the lawyer Clarence Darrow an opportunity to expose biblical literalism as foolish and harmful, principally through the answers he elicited from William Jennings Bryan, a champion of Christian values and leading opponent of evolution in schools. Most neutral observers declared the trial a draw. In 1960 the Scopes trial was turned into a popular film, *Inherit the Wind*, after which millions of Americans abandoned religious opposition to evolutionary theory.

The rise of similar creationist ideas today can perhaps partly be explained by the securities it offers in an increasingly turbulent world, fed by frustration at the growing divide between experts and populace, and a dislike of science performed behind closed doors. Deriving mostly from the prolific writings in the 1930s of the Seventh Day Adventist science teacher George McCready Price and revived in the 1960s by Henry Morris, a Southern Baptist preacher, 'young earth' creationists, 'flood geologists' and other believers in the literal truth of the Bible assert that the earth is less than ten thousand years old and that the fossil record was laid down all at once during Noah's flood. As Morris indicated in his *Genesis Flood* (1961), the Bible provides insufficient time for any kind of evolution. Morris's views are today promoted by the Institute for Creation Research in San Diego, from whence he and his followers denounce Darwin and offer a scientific-sounding alternative called creation science, publicized widely through school textbooks, brochures and revivalist meetings, and apparently supported by scientific 'facts' such as the finding of pieces of the Ark. Much of the distribution of information nowadays is electronic, for creationists have capitalized on the power of the internet to press home purported flaws in Darwin's reasoning and undermine modern Darwinism – a promotional device that reaches far more people than the academic profession's densely worded publications. Even though religious instruction is barred from American public schools, creationists make a constitutional case for including creation science on the American school science syllabus. Darwin's theory is only a theory, they say. Creation theory is claimed to be equally valid.

Influenced by the 'religious right' of Ronald Reagan's America, in 1981 the states of Louisiana and Arkansas passed bills to enforce 'equal-time' treatment in schools. Once again the American Civil Liberties Union brought an action against the Arkansas Board of Education that went all the way to the Supreme Court. Steven Jay Gould, one of the scientists called to serve as an expert witness for Darwinism in that case, felt himself almost to be appearing in a replay of the original 1925 court scenes in Dayton. His deposition makes fascinating reading. In a magazine article afterwards Gould reflected on the mismatch of definitions that occurred between judge and scientist:

> We define evolution, using Darwin's phrase, as 'descent with modification' from prior living things. Our documentation of life's evolutionary tree records one of science's greatest triumphs, a profoundly liberating discovery on the oldest maxim that truth can make us free. We have made this discovery by recognizing what can be answered and what must be left alone. If Justice Scalia heeded our definitions and our practices, he would understand why creationism cannot qualify as science. He would also, by the way, sense the excitement of evolution and its evidence; no person of substance could be unmoved by something so interesting.[1]

The court's final decision, made in 1987, was to ban the teaching of any creation science in publicly financed schools in Arkansas on the grounds that creationism was a religious concept and not a scientific one. Daunted but not neutered, many creationists have since gone on to establish independent Christian schools and colleges where creation science can be taught.

Today, across the United States, heated debates and lawsuits reflect rising support for the provision of alternatives to evolution in the state educational system. The Kansas Board of Education, for example, in August 1999, decided to make evolution optional in the criteria it issues for science teaching. Hence evolution is no longer covered in standardized tests for Kansas schoolchildren. Kentucky has deleted the word 'evolution' and substituted 'change over time'. These shifts in public opinion deeply worry scientists. Certainly, many scientists feel that an understanding of religious traditions has a relevant place in every child's education, not least in lessons on history and the development of diverse modern societies. Yet this is different from advocating a devotional standpoint as a real truth in science classes.

Even though the concept of the separation of Church and State lies at the heart of the American constitution, the United States is a uniquely Protestant country where the Bible still plays a crucial role. Partisans for a new variant, called Intelligent Design, argue persuasively for presenting this as an alternative to Darwinism in school classrooms. Intelligent Design does not generally refute evolution but suggests that some biological processes are far too complex to have originated in the step-by-step manner proposed by Darwin. Recalling many of the controversies immediately following publication of the *Origin of Species*, the biochemist Michael J. Behe proposes in his book *Darwin's Black Box* (1996) that protein reactions must have been designed by a superior intelligence. This is basically the old argument as put forward by William Paley or Asa Gray, brought up to date with new examples.

The new millennium has consequently begun with Westerners as divided as ever over the implications of a natural origin of species. Despite these challenges, the modern synthesis stands firm at the heart of biological science. No biologist would dream of disregarding the evidence. As Theodore Dobzhansky said in the 1960s, 'nothing in biology makes sense except in the light of evolution'.

History seldom tells of simple triumphant advances, but it can tell of the extraordinary impact of a single book. While many of the ideas and themes addressed by Darwin in 1859 were not new, and his writing style was mild in the extreme, the *Origin of Species* was clearly a major publishing event that spectacularly altered the nature of discussion on the question of origins. This interplay between one man, one book, and the diverse social, religious, intellectual and national circumstances of his audiences and the broader currents of historical change is what made Darwin's *Origin* such a remarkable phenomenon in its own day and which continues to absorb and instruct modern readers. Old texts are frequently remade by new forms of attention, and it appears that Darwin's *Origin* was both resilient in the survival of its main proposals and malleable in the hands of its devotees. His book can therefore be seen, not as a solitary voice deliberately defying the traditions of the Church or the moral values of society, but as one of the hubs of transformation in Western thought.

NOTES

Chapter 1

1 Nora Barlow (ed.), *The Autobiography of Charles Darwin, 1809–1882, with Original Omissions Restored*, London, Collins, 1958, p. 57.
2 Ibid., p. 60.
3 Ibid., p. 59.
4 F. H. Burkhardt and S. Smith et al. (eds.), *The Correspondence of Charles Darwin*, 14 vols., Cambridge, Cambridge University Press, 1985– , vol. I, p. 129.
5 Ibid., vol. I, p. 133.
6 *Autobiography*, 1958, p. 76.
7 Ibid., p. 78.
8 Ibid., p. 79.
9 Nora Barlow (ed.), 'Darwin's Ornithological Notes', *Bulletin of the British Museum, (Natural History) Historical series* 2, pp. 201–78, p. 262.
10 *Autobiography*, 1958, p. 80.
11 Richard Darwin Keynes (ed.), *Charles Darwin's* Beagle *Diary*, Cambridge, Cambridge University Press, 1988, p. 122.
12 *Autobiography*, 1958, p. 101.

13 *Correspondence*, 1985– , vol. III, p. 55.

14 Ibid., vol. I, p. 312.

Chapter 2

1 *Autobiography*, 1958, p. 100.

2 Paul H. Barrett et al. (eds.), *Charles Darwin's Notebooks, 1836–1844: Geology, Transmutation of Species, Metaphysical enquiries*, Cambridge, Cambridge University Press, 1987, Notebook B, pp. 63, 72.

3 *Notebooks*, 1987, Notebook C, p. 196.

4 *Autobiography*, 1958, p. 120.

5 *Correspondence*, 1985– , vol. II, p. 123.

6 Ibid., vol. II, p. 172.

7 Ibid., vol. III, p. 43.

8 Ibid., vol. III, p. 108.

9 *Autobiography*, 1958, p. 120.

10 *Correspondence*, 1985– , vol. VI, p. 335.

Chapter 3

1 *Correspondence*, 1985– , vol. VII, p. 118.

2 Francis Darwin, *The Life and Letters of Charles Darwin*, 3 vols, London, 1887, vol. I, p. 155.

3 Charles Darwin (1859), *On the Origin of Species. A facsimile of the First Edition with an Introduction by Ernst Mayr*, Cambridge, Mass., Harvard University Press, 1959, pp. 171, 188.

4 *On the Origin of Species*, 1859 edn, p. 31.

5 *Correspondence*, 1985– , vol. VII, p. 274.

6 *On the Origin of Species*, 1859 edn, p. 75.

7 Ibid., p. 84.

8 Ibid., p. 112.

9 *Correspondence*, 1985– , vol. VII, p. 265.

10 *Correspondence*, 1985– , vol. VIII, p. 75.

11 *On the Origin of Species*, 1859 edn, p. 488.

12 *On the Origin of Species*, 1860 edn, p. 484.

13 *On the Origin of Species*, 1859 edn, pp. 485–6.

14 *On the Origin of Species*, 1859 edn, p. 490.

15 *Autobiography*, 1958, p. 137.

16 *Correspondence*, 1985– , vol. VI, p. 178.

17 *Correspondence*, 1985– , vol. VII, pp. 324, 328.

Chapter 4

1 *Correspondence*, 1985– , vol. XI, p. 231.

2 *Correspondence*, 1985– , vol. VIII, p. 405.

3 'Agnosticism', was an essay first published by Huxley in the *Westminster Review* in 1889. It was reprinted many times afterwards. It is most easily accessed in Huxley's *Collected Essays* (*1893–1894*), vol. 5, p. 246.

4 *Westminster Review*, 1860, vol. 17, p. 556.

5 John Stuart Mill, *A System of Logic, Ratiocinative and Inductive, being a Connected view of the Principles of Evidence, and the Methods of Scientific Induction*, 5th edn, 2 vols., London, 1862, vol. II, p. 18.

6 Karl Pearson, *The Grammar of Science*, London, 1892, p. 369.

7 Charles Darwin (1871), *The Descent of Man, and Selection in Relation to Sex*, 2 vols., facsimile of the first edition, introduced by John Tyler Bonner and Robert M. May, Princeton, NJ, Princeton University Press, 1981, vol. II, pp. 368–9.

8 Ibid., vol. I, p. 57.

9 Ibid., vol. I, pp. 206–7.

Chapter 5

1 Steven Jay Gould, *Natural History*, October 1987, vol. 96, pp. 14–21.

SOURCES AND FURTHER READING

Barlow, Nora (ed.), *The Autobiography of Charles Darwin, 1809–1882, with Original Omissions restored*, London, Collins, 1958

Barrett, Paul H. et al. (eds.), *Charles Darwin's Notebooks, 1836–1844: Geology, Transmutation of Species, Metaphysical Enquiries*, Cambridge, Cambridge University Press, 1987

Bowler, Peter J., *Evolution, the History of an Idea*, 3rd edn, Berkeley, University of California Press, 2003

Brooke, John H., *Science and Religion, Some Historical Perspectives*, Cambridge, Cambridge University Press, 1991

Browne, Janet, *Charles Darwin, Voyaging*, New York, Knopf, 1995

Browne, Janet, *Charles Darwin, The Power of Place*, New York, Knopf, 2002

Browne, Janet and Neve, Michael (eds.), *Charles Darwin: Voyage of the Beagle*, London, Penguin Books, 1989

Burkhardt, F. H. and Smith, S. et al. (eds.), *The Correspondence of Charles Darwin*, 14 vols., Cambridge, Cambridge University Press, 1985–

Burrow, John, *Evolution and Society, A Study in Victorian Social Theory*, Cambridge, Cambridge University Press, 1966

Darwin, Charles (1859), *On the Origin of Species. A facsimile of the First Edition with an Introduction by Ernst Mayr*, Cambridge, Mass., Harvard University Press, 1959

Darwin, Charles (1871), *The Descent of Man, and Selection in Relation to Sex*, 2 vols., facsimile of the first edition, introduced by John Tyler Bonner and Robert M. May, Princeton, NJ, Princeton University Press, 1981

Darwin, Francis (ed.), *The Life and Letters of Charles Darwin*, 3 vols, London, John Murray, 1887

Darwin, Francis and Seward, A. C. (eds.), *More Letters of Charles Darwin: A Record of his Work in a Series of Hitherto Unpublished Letters*, 2 vols., London, John Murray, 1903

Dawkins, Richard, *The Selfish Gene*, Oxford, Oxford University Press, 1976

Ellegard, Alvar, *Darwin and the General Reader, the Reception of Darwin's Theory of Evolution in the British Periodical Press, 1859–1872*, reprint edn, Chicago, University of Chicago Press, 1990

Freeman, Richard, *The Works of Charles Darwin: An Annotated Bibliographical Handlist*, 2nd edn, Folkestone, Dawson Archon Books, 1977

Gould, Steven J., *The Mismeasure of Man*, revised edn, New York, W. W. Norton, 1996

Greene, John C., *The Death of Adam: Evolution and its Impact on Western Thought*, revised edn, Ames, Iowa State University Press, 1996

Healey, Edna, *Emma Darwin: The Inspirational Wife of a Genius*, London, Headline, 2001

Hodge, Jonathan, and Radick, Gregory (eds.), *The Cambridge Companion to Darwin*, Cambridge, Cambridge University Press, 2003

Hofstadter, Richard, *Social Darwinism in American Thought*, revised edn, Boston, Beacon Press, 1992

Huxley, T. H., *Collected Essays (1893–1894)*, 10 volumes, reprint edn, Georg Olms Verlag, Hildeheim and New York, 1970.

Jay, Mike and Neve, Michael (eds.), *1900 : a Fin-de-Siècle Reader*, London, Penguin, 1999

Kevles, Daniel J., *In the Name of Eugenics: Genetics and the Uses of Human Heredity*, revised edn, Cambridge, Mass., Harvard University Press, 1995

Keynes, Richard Darwin (ed.), *Charles Darwin's* Beagle *Diary*, Cambridge, Cambridge University Press, 1988

Kohn, David (ed.), *The Darwinian Heritage*, Princeton NJ, Princeton University Press in association with Nova Pacifica, 1985

Larson, Edward J., *Summer for the Gods: The Scopes Trial and America's Continuing Debate over Science and Religion*, Cambridge, Mass., Harvard University Press, 1997

Larson, Edward J., *Evolution: The Remarkable History of a Scientific Theory*, New York, Random House, 2004

Mill, John Stuart, *A System of Logic, Ratiocinative and Inductive, being a Connected view of the Principles of Evidence, and the Methods of Scientific Induction*, 5th edn, 2 vols., London, 1862

Pearson, Karl, *The Grammar of Science*, London, 1892

Peckham, Morse, *The Origin of Species: A Variorum Text*, Philadelphia, University of Pennsylvania Press, 1959

Ridley, Matt, *Genome: The Autobiography of a Species in 23 Chapters*, London, Fourth Estate, 1999

Ruse, Michael, *The Darwinian Revolution, Science Red in Tooth and Claw*, 2nd edn, Chicago, University of Chicago Press, 1999

Wilson, Andrew N., *God's Funeral*, London, John Murray, 1999

Young, Robert M., *Darwin's Metaphor, Nature's Place in Victorian Culture*, Cambridge, Cambridge University Press, 1985

INDEX

Note: The abbreviation CD refers to Charles Darwin.

Index compiled by Meg Davies
(Registered indexer, Society of Indexers)